The
TOLTEC
ART
of
LIFE AND
DEATH

The
TOLTEC
ART
of
LIFE AND
DEATH

A Story of Discovery

DON MIGUEL RUIZ
and BARBARA EMRYS

HARPER**ELIXIR**

HARPER**ELIXIR**

THE TOLTEC ART OF LIFE AND DEATH. Copyright © 2015 by Miguel Ruiz and Barbara Emrys. All rights reserved. Printed in the United States of America. No part of this book may be used or reproduced in any manner whatsoever without written permission except in the case of brief quotations embodied in critical articles and reviews. For information address HarperCollins Publishers, 195 Broadway, New York, NY 10007.

HarperCollins books may be purchased for educational, business, or sales promotional use. For information please e-mail the Special Markets Department at SPsales@harpercollins.com.

HarperCollins website: http://www.harpercollins.com

FIRST EDITION

Designed by Ralph Fowler

Library of Congress Cataloging-in-Publication Data is available upon request.

ISBN 978–0–06–239092–9
ISBN 978–0–06–243696–2 (BAM)
ISBN 978–0–06–243445–6 (BN)

15 16 17 18 19   RRD(H)   10 9 8 7 6 5 4 3 2 1

I dedicate this book with all my love and gratitude to the young woman who left her physical body in the month of October 2010 and donated her heart to me. Thanks to her generosity and to the generosity of her family, I have been able to travel to cities around the world, bringing my message of love, awareness, and joy to many people. It is because of her that I was also able to create this book with Barbara Emrys, whose imagination and artistry bring the story of don Miguel Ruiz to life within these pages.

To all the hospital personnel who have treated me since my heart attack, during my subsequent heart transplant, and up to this date, I offer my deepest gratitude.

I also dedicate this wonderful story to my sons, my daughters-in-law, and my entire family, all of whom I love so much. This is also for my readers, whose growing awareness over the last fifteen years has encouraged me to deliver my message in inventive and exciting ways. It is clear to me that their love for this wisdom has made the world a happier place to live.

Preface

This book recounts the events of my life. Unlike my previous writings, it merges the power of imagination with the teachings of Toltec wisdom. It tells the story of a mystic dream I experienced several years ago, during the nine weeks of a medically induced coma that followed my heart attack.

At the time of our death, it is said, a lifetime of memories will flash before our eyes. Something similar happened to me while my body desperately struggled to stay alive and my mind expanded toward the infinite.

You could say that during those long weeks I dreamed my legacy. A personal legacy is the compilation of all the experiences in our life. It is the sum of all our actions, all our reactions, all our emotions and feelings. It is what we give to those who remain, after we leave our physical body. A legacy is all that we are, the totality of ourselves. Through the memories others have of us, our legacy is determined . . . and the more authentic we are, the brighter that legacy will be.

I was inspired to create this book as a gift to my sons, to my students, and to all those whose love helped bring me back to life. To my children, my family, my friends and lovers, I give my memories and my unconditional love. To those who wish to learn from my words, I offer the experience of my life. My enduring love of the world is my gift to this beautiful planet. The authenticity of my awareness is my gift to humanity.

Our waking lives, like our sleeping dreams, are works of art. This book is an artistic piece of storytelling about very real interactions between me and my mother, doña Sarita, a well-respected healer in San Diego and my teacher and guide through much of my life. From the moment of my heart attack on February 28, 2002, she did everything she could to keep my body from dying. Using all the power of her faith, she gathered her children and apprentices to perform a series of ceremonies on my behalf. She worked tirelessly, day and night, to bring me back to health and consciousness. She was determined that I return to my body and give it life again. On many occasions she went into a trance, or deep meditation, with the intention of entering my dream and demanding that I reject death.

Those excursions into my dream state are the basis for this book. When my mother confronts me there, I send her to talk to the main character of my story, which is my own knowledge. In this fantasy, knowledge is depicted as a mysterious creature called Lala. You could say she is the embodiment of everything I believe and everything that gives shape to my story—just as your knowledge helps you create the story of your life.

Many wonderful characters bring energy and life to this story. Each one reflects me and each one contributes to my healing in a singular way. Although some of their names and some of their exchanges with me have been fictionalized, all of these characters represent actual friends, students, and family members. Some are dead and some still live and laugh with me, but all of them have enriched my world. My love for each of them is strong, and my gratitude for the role they played in my life and in my recovery is boundless.

It may seem that our experiences—yours and mine—are very different. Your main character is different from mine, and your secondary characters probably don't seem like the people in my story. While we may seem different, you are an essential part of the dream of humanity, as I am. You have searched for truth through symbols, as I have. You are knowledge, seeking to redeem itself, as I was. You are your own savior, and you are pure potential in action. God represents the truth of you, and the truth will set you free.

Let this book help you understand these things. Listen, see, and dare to change your own world, a world made of thoughts and automatic responses. Allow the events of my life to inspire new insights about your own dream and its current challenges. A good student makes the most of whatever information becomes available, and as my story demonstrates, life provides all the information we need.

With all my love and respect,

—Miguel Angel Ruiz

Glossary

Awareness: The ability to see things as they are.

Death: Matter; the absence of life.

Dios/Diosa: God/Goddess

Don/Doña: Titles of respect in the Spanish language (similar to sir, lady).

A dream: The mind's reflection of our perception.

The dream of the planet: The collective reality of the human species.

Dreamer: Someone who knows that he or she is dreaming all the time.

Energy: The eternal supreme power, the only thing that really exists.

Evil: The result of believing in lies. Evil actions intensify according to how distorted the lie is and how great the level of fanaticism.

Faith: Believing 100 percent, without a doubt.

God: The eternal supreme power, the only thing that really exists.

Heaven: A story in our mind that results in happiness.

Hell: A story in our mind that results in drama.

Intent: The message of energy that gives direction to light, creating matter and disintegrating matter. Intent travels at the center of light, with quanta revolving around it. Intent is life itself.

Knowledge: Agreements made between humans about the nature of reality. Knowledge is communicated through symbols, such as words and numbers, phrases and formulas.

Lies: Distortions of the truth within the human mind.

Life: The creative force of God, or energy, that manifests matter.

Light: Life's messenger, and its first manifestation.

Love: The aspect of energy that manifests as the totality of all vibrations, moving matter and recording information into matter. Matter perceives and reflects it, and reacts with the complete range of emotions.

Magic: The creative aspect of energy.

Matter: The finite manifestation of infinite life.

Mind: A virtual reality, created by the reflection (in the brain) of everything that the brain perceives.

Mitote (mee-'toe-tay): The ongoing conversation in our head, which sounds as if a thousand people are talking at once and nobody is listening.

Nagual: The Nahuatl word referring to the force that moves matter.

Nagual man/woman: A person who knows himself/herself as the force that moves matter; an immortal.

Nahuatl: The language of Aztecs.

Power: The potential to create.

Shaman: In all cultures, a medicine man or woman.

Soul: The force of life that holds a universe (matter) together (e.g., the universe of the human body). Every component recognizes itself as part of that universe.

Story: An explanation of a dream.

Teotihuacan (Teo): An ancient city of Mexico, which flourished from 200 BC to AD 500. Its excavated temples and pyramids are located approximately thirty miles northeast of Mexico City.

Toltec: A Nahuatl word meaning artist.

Tonal: Matter

Truth: That which is real; another word for God and energy. Truth existed long before humanity and will exist long after humanity.

Wisdom: The ability to react correctly to every event; common sense.

Cast of Characters

Don Miguel Ruiz: The main character of his story.

Mother Sarita: Don Miguel's mother and teacher.

Lala: Knowledge

José Luis: Don Miguel's father and Sarita's husband.

Don Leonardo: Don Miguel's grandfather and Sarita's father.

Don Eziquio (*ay-'see-kee-o*): Don Miguel's great-grandfather and don Leonardo's father.

Gandara: Don Eziquio's friend.

Memín: Don Miguel's brother.

Jaime ('*hī-may*): Don Miguel's brother, closest in age.

Maria: Don Miguel's wife and the mother of his children.

Dhara: Don Miguel's apprentice and romantic partner.

Emma: Don Miguel's apprentice and romantic partner.

Miguel, Jr. (Mike, Miguelito): Don Miguel's eldest son.

José: Don Miguel's second son.

Leo: Don Miguel's youngest son.

The
TOLTEC
ART
of
LIFE AND
DEATH

Prologue

I PULL AT THE BEDSHEETS, TIGHTENED NOW AROUND my ankles. I reach for the phone, dial blindly, and then someone is talking to me. A woman is asking me who I am, where I am. It seems unlikely I will remember the answer to either of those questions before speech leaves me forever. I try to sit up, but roll from the tangle of sheets instead and tumble to the floor. The pain slips away mercifully, only to come back again in furious stabs. I can hear my mother calling me, shouting my name. I can hear the voices of strangers and the wailing of sirens as consciousness slips between the rise and fall of incongruous sounds. There will be sweet goodbyes, as a new dream begins to rise in place of the old, but all I recognize in this moment is the distant sobbing of women.

So many women are crying. They cry for a son, a lover, a father, and a guide. They cry for me, for themselves, and for promises that were never made. Like all humans, they cry for the redemption of a word. They cry for Love, the fallen angel, when they need only look, listen, and feel the force of it pounding like music from their own wondrous bodies.

Today, I woke before daylight to an invitation from Death. Like my Aztec ancestors, I welcome it with the gratitude of a warrior who has fought well and wishes for a safe homecoming . . . and a long rest. On some distant horizon I can feel the glow of approaching dawn. My skin warms to it. My eyes lift to see mist dissolving into star-fire, and I know I'll soon see the way home, out of this dark night. My adversaries have come and gone, vanquished by love. They fought relentlessly within the hallways of the human mind, that splendid battlefield. There will be others like me, eager to lift their swords against a million lies, but the war is over for Miguel Ruiz.

Just moments before, as I slept and dreamed, I had a vision of another warrior, a young man from an ancient time, standing among the foothills of a sacred mountain and watching over his beloved valley. He stood under the faintest starlight, gazing at the lake that curved protectively around Tenochtitlan, the home of his people and mine. In the dream, the great valley was veiled in mist. Slowly, dimly, predawn fires began to twinkle as his village came gradually awake. The young man's heart was beating loudly, as mine is now. His nostrils tested the night air, and his skin tingled in response to wind shifts. Lowering himself carefully to one knee, he lifted his bow and held it high. The fingers of his right hand touched the feathers of an arrow blessed by smoke from a sacred fire. He would not fail his people when the attack came. He would not fail his family, nor the memory of the ancient Toltec people. He would not fail himself.

This was the most dangerous time, the hour when morning had not yet imagined itself and good struggled against evil in the predawn gloom. The young warrior blinked his eyes once, then again, and steadied his arm. As I dreamed with him, it seemed I could feel pebbles shift under one sandaled foot, bite into the flesh of his knee. I could feel the mist grip the man's ankles and tighten its chilly hold on his bare arms and thighs. I could feel it licking at the back of his neck and his painted brow. Together, we glanced toward the sky. The world above him—an array of stars within a field of mystery—mirrored his perfect body. Seeing this, he whispered a prayer and steadied his breath. His body relaxed. His attention moved back to the valley below, where the mist had begun to disperse and the waters of his ancestral lake curved between dark hills like the jeweled fingers of a goddess. He steadied the bow. The eagle feather in his hair danced gracefully in the rising wind. His back was straight and his belly relaxed. His dark skin glowed radiantly bronze in the approaching sunrise.

His people would be grateful to him now. He imagined some of them peering out of their doorways and sensing the threat that lay beyond the fog. He looked toward the lakeside village as if he might see his father gazing back at him where he knelt quietly and alone—one brave soldier empowered by the strength of the fiery mountain. He felt his father's pride, and the pride of the ancestors. There was so much to feel in that empty moment between the start of things and the end of things. Light would soon burst over the eastern rim of mountains and destiny would rise up shouting behind it. There were victories lying in wait. Revelations loomed just beyond this present uncertainty. With the breath of his ancestors on his cheek, and the cool touch of their hands upon his back, the warrior composed himself again, one sandal digging into the rocks and eyes staring down the shaft of his warrior's arrow. He was ready. . . .

And now, as the shock of pain startles me from my dream, I see that it is my time to join the warriors of antiquity. As I once stalked truth, eternity stalks me now. Sunrise thunders along the eastern ridge, and destiny is riding in its wake. With the breath of my ancestors on my cheek, and the cool touch of their hands at my back, I wait for Death's greeting with a welcoming smile.

I, too, am ready.

THE OLD WOMAN MUTTERED TO HERSELF AS HER
feet shuffled along the surface of the dry, cracked terrain. Her slippers scratched the dirt, kicking silky
clouds of dust into the wavering air. She held a large bag in one
hand and clutched her shawl around her shoulders with the
other. The beat of her labored footsteps was the only sound,
a slow and plodding sound that never hesitated. She walked
on. There was no path to speak of, but she didn't need one.
She knew where she was going. She was following the traces of
something invisible to her, but unmistakable. She was following the instincts of a mother searching for her son.

For weeks now she had felt the chilling fear a mother feels at
the possibility of losing her child. Somewhere in the world she
had just left, her thirteenth child was slipping away—not from
her sight, for she knew he lay silent and pale in a hospital bed.
He was slipping steadily away from her senses. She could no

longer feel the life-current of him. She could no longer speak to him in the wordless ways that they had shared for almost fifty years. As the force of life weakened in him, so did his ties to the world of matter and thought. There was very little time left, she knew. His heart had failed, his body was dying, and the doctors were poised to give up the fight. What else could she do but journey into this timeless place where his presence had gone, and seek him out? She would find her youngest son, the soul of her soul, and she would bring him home.

Beyond her fragile form there stretched a vast landscape of sand and rock and all manner of lifeless things. There was no color, save for billowing clouds of slate blue that swarmed above her soundlessly. Lightning seared the depthless heavens, blinding her in jagged rhythms . . . but this storm was made of dreams. This was a storm born of feeling and wonder, and such things would not slow her progress.

Sarita continued on, the sound of her breath echoing into the silence. Her pulse quickened and her chest heaved, as if her exertions were real. Perhaps they were. She had never attempted such a journey before. She had not known what to expect, or what cost her body would have to pay. As she walked on, she willed herself to relax. She would not succumb to fear. She was old; it was true. She had recently celebrated her ninety-second birthday, but she was not ready to leave the world of matter and meaning. She was not ready, and therefore *he* was not ready. Her son would not be permitted to die while she still had the strength to fight for him. She took a quick breath and allowed a smile to wash the strain from her features. Yes, she had the strength. In this peculiar space between here and there, her love would triumph. Encouraged, she set her bag down for a moment and straightened her shoulders, gathering the ends of her shawl in a loose knot at her neck. She was wearing a night-gown made of thin cotton. The windless cold seeped through it easily, chilling her flesh. No matter, she thought. There was no turning back now. Her senses might fail to recognize him, but her heart would not. Scanning the landscape once more, she picked up the heavy bag with the other hand and resolutely shuffled on.

It was a nylon shopping bag, the kind that she would have taken to market in those cool early mornings in Guadalajara, during the days just before her youngest had been born. It showed a portrait of the Virgin on the outside, printed in bright colors, and within it were many items blessed by her own prayers and intent. She gave the bag a gentle shake, as if to reassure herself of her mission, and thought of those days so long ago, just before the birth of her thirteenth child, when all of life seemed reassuring. It had been a sweet time: she was forty-three, still beautiful, and wedded to a handsome young man to whom she had already given three sons. He had married her right out of school, in spite of her age and her nine children by a previous marriage. He had married her against the wishes of his family. He had married her, some said, because she worked her wicked magic on him. Well, there would always be those who were skeptical. They had married out of love, pure and simple. From love, four healthy sons were born.

The old woman slowed her pace, then stopped. The storm still flashed and billowed around her, but its eerie silence was gone. Now, beyond the muffled sounds of her breathing, there was something else in the air. Where there should have been thunder, there was now music, building in the distance like a growling wind. He must be near, she thought. She stood where she was, listening, until it became clear that a particular song was playing, rising from the horizon to meet the sky's fury. It was music she recognized from a time long ago. She could hear her son singing to music like this as a boy, his little fingers moving along the strings of an imaginary guitar while he mouthed senseless syllables and shook his whole body to the rhythm of it, just as he had seen his older brothers do. What had he called this sound? What . . . ? Oh, yes.

"It's rock-and-roll, *Mamá*!" she remembered him shouting. "The music of life!"

Yes, a rock-and-roll song was playing in his head even now. That was the sound that raced along the lightning bolts in this blackening sky and whipped like cyclone winds through her gray hair, even when everything around her was still. Her senses had not failed her. She could feel his mind now, and hear his immense and eternal heart reverberating with joy. He was close.

Setting down the shopping bag again, she wrapped her woven shawl more tightly around her. She was dressed for bed, wearing what she'd had on when everyone had arrived at the house to join her in ceremony. In some distant corner of her consciousness, she could hear those guests, too—her children, her grandchildren, her students, and her friends. They had come at her request—for the obvious reason that no child or grandchild, no apprentice or assistant, ever refused Mother Sarita. They had come in quiet resignation—bringing gourds and drums, lighting candles, and burning sage. They had come to sing, to pray, to plead. They had come to bring him back, the thirteenth son of a woman who could not be ignored. They had come as the ancestors would come, to do the job of spiritual warriors.

On this night, with so much at stake, Sarita had been transported from the circle of the faithful in her living room to a world that existed only in imagination. She had trespassed into the mind of another. She was willing to pay the price for that at some other time, but for now she must keep going. For now she must walk without apology into her son's dream, and she must bring him back—dragging him by an insolent ear, if she had to. Certainly, she had done it many times before.

She shook her head as she remembered the child he once had been. She remembered those black eyes full of humor and mischief, and the little hands that had reached for her face with love when she was tired or touched by sadness. There was nothing—not even death—that would keep her from him. There was no logic that could undo her need for him, not even his logic. In her ninety-two years, Sarita had experienced all the joys and sorrows of being thirteen times a mother. She had survived the deaths of two of her children before this. She had lost husbands, sisters, brothers—but there was enough life in her still to fight one last time for what she loved. Picking up the bag again, she shook a little ethereal dust from the image of the Virgin Guadalupe and searched the landscape. She sniffed the air for some other sign, hesitated, and then turned around. Something had caught her attention, something that could not yet be seen. She would change course. She must follow her intuition—and the music.

The music grew louder with every painstaking step she took. It seemed to vibrate from ground and sky at once, pulsing to a loud beat . . . perhaps to the beat of the drums in her living room. She thanked God silently for obedient children and continued walking, her feet moving heavily through a thick spray of illuminated dust. Beyond the near horizon, she could see Earth rising over the rim of this vacant dream, blazing with a spirited light. She caught her breath. In the darkening sky of storm and shimmering heat, she could see something silhouetted against Earth's brilliance. A tree loomed in the distance! Its heavy limbs seemed to undulate with erotic pleasure, causing green leaves to quiver and shine. Sarita marveled at the sight of something so full and fertile in a land of such vast emptiness.

Miguel . . . she whispered. In any dream where there was color and life, there would be her son. He used to say that fun followed him everywhere. Well, this was fun. This was magic. Wherever he was, there would be a celebration—of that she was certain. She walked on toward the tree, the music growing louder. The walk might have taken a lifetime, or a minute, or no time at all. She was aware only that her heart was beating to a lively tune while she walked. She must have come a long way, whatever the time, for the massive tree spread before her now— tall, wide, and graceful. Its limbs stretched in all directions, as if beckoning the universe into a huge, benevolent embrace. Sarita hesitated by a root that jutted out of star-silt, and peered up into what looked like a galaxy of suspended fruit twinkling in the unworldly light. As she gazed in wonder, her eyes fell on the one she had come to find. On the lowest limb of the gigantic tree, almost hidden among the dancing shadows and the thousand sparkling leaves, sat her son.

Miguel Ruiz was lounging against the trunk of the tree in his hospital gown, quietly munching on an apple. Seeing her now, his eyes brightened and he waved enthusiastically for her to come closer. His mother edged toward the tree, choosing her steps carefully through the enormous tangle of roots, until she was standing by the limb that supported him. It swooped low along the ground, making it possible for her to look directly into his eyes.

"Sarita!" he exclaimed, wiping juice from his lips with the tip of his thumb. "You've joined me! Good!" As she was about to speak, Miguel turned his whole body in the direction of the improbable horizon. "Do you see what I am seeing, *Mamá*?" Miguel pointed enthusiastically at the vision of Earth and all her exquisite colors. Sarita caught a glimpse of her son's bare bottom as the back of his gown fell open. She was tempted to spank him right there, grown man that he was, but he was anxiously calling for her attention.

"Sarita, look!"

From where she now stood, she could see the planet floating beyond the bending branches of the giant tree. It shone bright and clear against a midnight sky, spinning slowly at the edge of the fantasy they occupied.

"*La tierra*," she said, sighing. "Where we both belong. It is time to stop this idiocy."

"Do you see them?" Miguel asked urgently. "All the moving lights?"

Frowning, the old woman peered through the branches again. This was not Earth as she remembered it. As the planet slowly turned, she could see waves of light burning bright, then lifting away and evanescing into space. The lights burned hot in some isolated places, and not in others. But wait . . . no. Some streamed over the entire globe. And even as little sparks rose and dissolved, more waves of light fell onto Earth like liquid dreams.

"Yes! Dreams!" her son exclaimed, as if he had followed her thoughts. "These are the dreams of men and women who change humanity. Small ones, bigger ones, and great, lasting ones. Dreams that begin and end, live, and then die."

"If they die, where do they go?" she asked, puzzled at the rising and falling of light, much like the bouncing waves of sound displayed on her grandson's stereo. "And where do they begin?"

"From creation—and back to creation!" he said with a laugh, taking another bite from the apple. "Do you see that bright one?" he marveled. "Wonderful! It feels like George, whose message is still remembered. So gentle a dream . . . do you see it?"

"George . . . ah, yes. He was your student. The very short one?"

"No, he was one of the Beatles, Sarita. And much taller than me."

Oh, yes. Now she remembered. *The Beatles*. The sound that had serenaded her to this spot was their sound, their music.

She was only now recovering from the throbbing noise in her head.

"Do you see my dream, Sarita?" Miguel shouted. "There! It shines in that area over there! And look! The threads of it are moving, getting brighter . . . everywhere! There! A yellow-gold—no, *red*-gold there. Wait!"

Sarita let the bag drop from her hands and gripped his shoulder. Miguel swung around to look at her, his face still glowing with joy.

"Your message is alive and growing, yes," she said. "There it is. We see it."

"Isn't it magnificent!" With that Miguel abandoned his apple, tossing it aside. It vanished as soon as it left his hand. He moved to observe the vision of a dreaming humanity more closely, but his mother's words distracted him, sounding stern and cheerless.

"We need Miguel to keep this dream alive. You are returning to me now," Sarita said in as strong a voice as her son had ever heard. "It is not your time to die."

"I'm already dead," her thirteenth child answered, smiling.

"You are not. The doctors are caring for you. We are praying for you. The ancestors are moving heaven and Earth for you."

Miguel twisted his face in mock despair, but his eyes still gleamed. "*Madre,* not the ancestors, please."

"Your heart is mended now, *m'ijo.* You have only to take a breath and come back to us. Come back!"

"You're talking about a heart that's damaged beyond repair, Sarita. My lungs have failed and my body is collapsing without me." He looked at her tenderly. "I'm a doctor, too, remember."

"You are a coward as well! Come back and finish what you began!"

"You know that I've given all I can."

"*Have* you?"

"Oh! Let me tell you about the sleeping dream I had before I got here!"

"Miguel."

"I was one of the warriors who guarded Tenochtitlan and the sacred lake. I was—well, of course I wasn't, but in a way I still am—that warrior. I could feel the fear and the urgency of the moment, the total surrender, and then it seemed that everything became starlight and space."

"Stop, Miguel! Your world is more than starlight and space. You have a home, and people who love you. More than that, you have me. You are my son, and you must return to me!"

"All of it is starlight and space—this world, that world, this mother and this son."

"You are not starlight and space. You are—"

"I am *exactly* that! Look at me!" With that, he disappeared among the twinkling orbs that danced before her eyes. There were only stars now, and the space between.

"Come back!" she shouted.

"Impossible," he replied, laughing, and she saw him again within the tree that seemed to come and go, straddling another limb, his bare legs swinging as he waved to her. "Stay with me, *Mamá*."

His mother's fear exploded into fury, and in that moment Miguel saw her transformed. The frail old woman who had come to him, wrapped in a shawl and shivering with cold, was an old woman no more. Before him, in the full sun of an eternal moment, stood a young and beautiful woman, naked but for the shawl that fell from her beautiful breasts and shoulders. She scowled at him, her hair caught in the wind that had risen in her anger. A fierce light shone over her, licking at her hair and skin like dragon fire.

"You are mine!" she raged. "How dare you leave! How dare you!"

"I haven't left you, my beloved," he replied gently, watching her with intense interest. "But the dream of Miguel is done. Game over."

"*Not* done! *Not* over!" she cried. "You can do much more—and you *will* do much more!" She turned her angry gaze toward the planet again, and pointed at the glittering lights. "Are you content to see your dream fade—here, right here before your eyes?"

Miguel, recognizing this voice, answered with a smile. "You can't move me, my love. My journey is endless, but my poor body won't go another mile."

"The body will do as you say. It always has! Come away from this place and return to me . . . to us!" In the far distance rose the sounds of his family—brothers and sons, their wives and children—as they chanted in a circle, calling for his return to the physical world. He knew they meant to help. He knew they followed his mother's will.

"I cannot," he said simply.

"You are mine!" she shouted.

"I never was."

Miguel looked into the eyes of his beloved and saw her beauty, her sorrow, and her worth. He heard the pleas of his mother, but could comprehend only the desperate cry of this one—who had been called many things in human storytelling. She represented humanity itself, a vibrant miracle trapped within its own spell. It was she who had lost the memory of paradise. It was she who had cast a shadow across sublime light. As he looked at her, remembering countless others who had said they loved him while they raged against themselves, his voice softened and he reached for her.

"Your temptations are strong—stronger even than your need for me." The touch of his hand on her bare arm cooled the fire in her eyes, and he began to see his mother, old again, and trembling from an unfelt chill. She gazed at him, her eyes softening, pleading.

"Don't worry yourself, Sarita," he soothed. "I am everything now."

"And what of me?" she asked, sounding like a child as she shivered in her nightgown, looking at him with wide, fearful eyes. "Do not leave me," she cried. "Do not abandon me to a world that does not include you."

"Miguel can't return. He's dead."

"The old ones sometimes brought the dead to life!" Her eyes flashed, and then she lowered her gaze self-consciously. "I will ask. They will know, *m'ijo*," she muttered.

"They would not bring back Miguel, your son, even if he agreed to it. He will be a spent dream, attempting to survive within a dying body."

"So . . . it might be done!" his mother exclaimed. The fire was in her eyes again and he felt the temptation that burned strong behind it.

"Sarita, do not ask this."

"I will have you back! I will, or—"

"Or what—or you'll die? Do it now! Come home with me!"

"I am not ready for this bleak surrender!"

"*Madre,* you don't listen."

"Come back, then, and make me hear you," she cried. "Come back and teach me what I would not learn."

Miguel sighed. She was using words to bend him, as she always had. It had never been easy to win an argument with her. Sarita had been his teacher, his patient master, and it was hard for him now not to respond as a student. He leaned heavily against the trunk of the tree and turned his attention to the great, glittering sphere that floated above the horizon, welcoming certain dreams and abandoning others.

"Your dream is fading already," Sarita pressed on, following his gaze. "Such a tragedy. Your sons are not strong enough without you; your apprentices are weak and selfish."

"It doesn't matter, Sarita. They are happier than they used to be. The world is happier." He turned back to her with a look of contentment.

"Who gave birth to you?" she snapped. "Who taught you, and trained you, and prepared you to seduce Mother Earth herself?"

"*Tu, Mamá,*" he answered quietly. He knew what was coming. It would be hard to say no to her, as it had been hard to say no to the rest of her kind. She counted on that.

"Obey your mother. Time is running out, and I will not return without you."

"And I ask you to join me, Sarita. There is nothing left for you but physical suffering. I would spare you that."

"Do not paint me as a victim!"

Miguel regarded her thoughtfully. She was not a victim. She was a woman who abhorred the ravages of age and would not willingly face the end alone. They had collaborated for fifty years now, like two children inventing games—games, in this case, that changed the dreams of human beings. In his absence, there would be no one like her left in the world . . . but did she understand the price his body would pay to come back? Could she imagine the extent of his physical pain? Something stirred in him, and he felt the force of his love begin to shift the dream. He looked into his mother's eyes and spoke to her, choosing his words carefully.

"If this body lives, *Madre,* it will need my presence; but it will also need something of the old structure."

"Was it not I who taught you about the human *form*?"

"There's no form left—no belief system."

"Such things can be retrieved!"

"Who was Miguel, Sarita? How can he be recovered, when there is no answer to that question? There are only memories to point the way. Memories lie, and the lies change with every telling. Memories may give direction, but never truth."

"They will give me *you*!"

Miguel looked at his mother, a vision of shifting moods and remembered phrases. She seemed real, warm, and so sweetly unassuming in her nightgown and slippers that he was tempted to change the conversation to everyday things. He wanted to tease her again, to make her laugh as he used to. He wanted to hear her calling him to breakfast, or casually gossiping about people he didn't know. He wanted to feel her fingertips on his forehead, over his heart, as she gave him her usual morning blessing. This was not an ordinary encounter, however. She had found him somewhere between life and death. She had found him because life had laid a path for her . . . and now, instead of yielding to this fragile dream, she was attempting to manage it.

What could he offer her as consolation for a lost son? How could he calm her fears as he once did? She was fighting him, and it appeared she would not stop. She seemed set for battle, even as she stood unsteadily before him, an old woman in a cotton gown and slippers. She would be the warrior, frail as she was, until it became obvious that there were no more wars to fight. What she hoped to win he could not say, but she was plainly determined.

Miguel offered her a smile. "You have a shopping bag, I see. Was it your intention to put me in it?"

"I might have!"

"It appears to be full already."

"Here!" she exclaimed, her voice raspy from all the talk. He noticed her renewed enthusiasm and let her talk. "I brought the usual tools of our trade! Perhaps we can do ceremony together . . . just as we used to. Prepare yourself, *m'ijo*. Make yourself pure, and bring the forces of life toward our task."

Miguel did nothing. He watched his mother patiently as she bent over her bag of treasures, one hand resting on his knee and his eyes shining with a curious light. He had been a shaman once and knew what was coming. The time was over for tricks, but how could he tell her that? The dream was over for Miguel, the main character of his story, but she would not listen. She would insist on having her son returned to her, even if he was a faintest copy of the truth, living within the most tenuous form.

Sarita began lifting items out of her shopping bag with pride and newfound enthusiasm. Could it be that she and her playmate of old were to invent yet another new game? Could fortune be on her side again? She felt the nearness of her ancestors and smiled. Out of the heavy bag she pulled a small drum and stood it on the ground, carefully placing a stick wrapped in ceremonial red ribbon on top of it. From a tiny pouch she shook out a collection of Aztec shards and lined them up neatly on the skin of the drum, adding to the arrangement a glorious eagle feather. That done, she stacked three gourds at the base of the drum, along with a pot containing charcoal and frankincense. Satisfied that she had laid the groundwork for all that was to come, she reached into the bag for her precious icons, and one by one she placed them on the limb of the tree.

"Now! We start with the Son of the Virgin, of course!" She balanced a small figurine of Jesus on the broad limb of the tree. It was a clay piece, daintily sculpted, showing the Lord holding a lamb. Next, she brought out the Virgin Mary, arms opened in an ascension pose. "There. Mother and Son united," Sarita said with satisfaction, then muttered a prayer.

Miguel watched in silence as she finished her prayer and hesitated, apparently unsure what to do next. Pursing her lips, she leaned over the bag again. After a few seconds of rummaging noisily, she straightened up, a brass statue of the Buddha sitting heavily in both her hands. She looked at her son, as if expecting a challenge.

"And why not?" she asked. "Is he so proud that he cannot come to the aid of a fellow teacher?"

"He is not proud, although he has good reason to be," said Miguel calmly, nodding his head toward the lights that flickered above him. "His message still moves the dream of humanity."

"Precisely so!" The old woman lifted the statue onto the tree, wedging it in the joint of two limbs. Closing her eyes, she mumbled another prayer, presumably to the ultimate bodhisattva himself. With another sigh of satisfaction, she reached into the bag again. This time she found a more delicate statue, wrapped in a silk cloth. It was a Chinese goddess, represented beautifully in pale jade. After a few seconds of consideration, she placed it beside the Virgin.

"A mother hears the cries of her children. She will answer." Sarita looked at the two women, standing gracefully under the light of the living world, and she smiled. "Yes, a mother answers."

Next came another brass figure—this one an elaborate version of the war goddess Kali. Miguel wondered how many households his mother had ransacked to fill her bag with fetishes. It was doubtful she knew the names of these goddesses, much less their significance.

"What do you think?" Sarita asked. "She seems like a fighter, but I don't want her to think that death is our objective."

"You may see that there are greater things to battle than death."

Sarita looked at her son as if seeking comprehension. He met her look, and she felt more confusion than comfort. Looking quickly away, she reached for the nylon bag and shook it. There was something left at the bottom. Grabbing it, she brought it out with a shrug and a sigh. It was his childhood plastic figure of Popeye, pipe in mouth and both biceps bulging. This she had found in his dresser drawer.

"Now we can talk!" exclaimed her son, laughing. "I am what I am!"

Sarita smiled with satisfaction. The meaning of this silly item eluded her, but she had been right to suspect that it would please him. She withdrew her withered hands and tugged on her cotton gown nervously. What else? Feeling around for a pocket, she withdrew a necklace: a silver chain holding a star of David. This she hung from a leafy twig, and gave it a spin. Then she took the gold crucifix from around her neck and draped it over the same twig. The two charms spun and gleamed in the surreal light, sending little sparks of fire into the upper branches of the tree. "Old gods, young gods. How are they different?" she whispered.

"Why bother with gods at all?" her son asked. "Why call on the saints and the ancestors? Why bring any of them to a conference between mother and son?"

"Because we need help."

"You need faith—but not in them."

"Then . . . in what?"

"Is it possible you're asking me this?"

"I have great faith in you, my lamb."

"Not in me. Faith in *you*. It's what brought you here, guided you to me. Faith is life itself, breathing through matter and moving us both."

"You are not moving at all."

"Am I not? Haven't I been moved already?" He gave his

mother a look of resignation, shaking his head. What more could he tell her?

"*M'ijo*," his mother said softly, clearly. "I will have you return to me, or I will die trying."

Yes, I see that, he reflected. Now, however, she was alive. Life still pushed through her, invigorating an old body with an unmistakable will. If she were to revitalize him, she would need that will to become even stronger, for he had slipped past her emotional reach. She would need total faith, which could come only from an awareness that presently eluded her. Yes, even Mother Sarita, sage and healer, had revelations waiting . . . and a journey ahead of her, too long postponed.

"You will not die today, Sarita," he stated at last. "Nor, apparently, will I."

He must take this chance to attend to her. His mother had always been ready to fight for him. She had always defended his right to be who he was and to achieve what he wanted. This time she was defending his right to live. As he saw the light come back into his mother's face, the face that had graced him over the years with a thousand expressions of love and pride, his imagination was set alight. He would give Sarita a mission, if she felt she needed one, and give the warrior one last battle to fight. While he still could, he would set her on a journey far more important than its intended destination.

"You say you will do anything?" her son asked.

"Yes!"

"Even if it means following instructions?"

Sarita could feel her heart beat faster. "My angel, in this peculiar world, you are the teacher," she said. "I will gladly take your instructions."

Okay, now who was teasing whom? Miguel thought wryly. Even a dying man had to laugh. And he was surely dying . . . the process had begun. He could see that Sarita had come to him as an impassioned force of life; and in a dream made of memory and waning desires, only life could stop that process.

"Not my instructions, *Madre*," he said, his smile brimming with love. "In my peculiar world, the outcome makes no difference. In someone else's world it is everything." He looked past her, to something in the distance.

"What do you—" she began. "Someone else?"

Sarita's eyes followed his gaze to a point along the far horizon. "What is this?" she asked. "Another tree?"

Far from this gleaming place they occupied, on another hill in a similar landscape, loomed an enormous tree. She hadn't noticed it until this moment. It was in every way the same as this one, the one that held her son on its noble branches. It was . . .

"A copy," he informed her.

"And who sits there? A copy of my son?"

"An impostor of another kind. The one who lives in that tree knows the science of illusion. Speak to that one, Mother."

Sarita looked across desolation to the tree in the distance. It was obscured by shadow, but radiant with color, as this one was. Nothing moved, however. Its leaves did not flutter, and nothing shone. Shadows did not play with flickering rays of light. There seemed to be no living thing among its branches. She was mesmerized. It took a deliberate act of will to look away and return her attention to her son, there in his Tree of Life, where he sat silhouetted against the brilliant colors of Earth.

"It is not more illusion I want. It is Miguel."

"Your journey begins there, Sarita," Miguel advised, taking another glance at the tree in the distance. Everything perceived was reflection, illusion. She would now have the chance to make her choices based on that awareness. "If you must know how to bring back your son, there lies your first instruction. As always, believe nothing you hear, but listen."

He plucked another apple from the branch above him and began polishing it on the hem of his hospital gown. He took a hearty bite, and as he began to chew, sweet juice streaming down his chin, he lifted his eyes to the black sky and grinned with profound delight at the vision of a planet blazing with dreams. His mother would prove herself adept, he had no doubt. Her awareness would grow with every challenge. She would put her considerable wisdom to use and consult the ancestors, as she always had. She would deal with the one who rules the world of reflections—a world he had left far behind—and, for a while at least, she would forget the pain that springs from a mother's intolerable fear. He winked at her cheerfully and readied himself to follow life, wherever it led.

Sarita smiled back, confident now as she felt the power of her intent moving time and circumstance forward. She must stay in her son's dream, no matter what. Here, she could persuade him. Here, he would feel the force of her will. In her mind, she had made her case well, and for now he was conceding. He was pointing the way to a solution, however dubious it appeared to her; and this was progress. She would indulge him, of course. She would try things his way . . . until his way became her way.

Sarita set her eyes on the horizon. No one could face what lay ahead but her, however many hours her family might spend on music and prayer. She turned from Miguel without another word, picking up her empty bag, and began walking again, this time toward whatever lurked in the shade of the great tree in the distance.

There was no wind. In this still landscape, canopied by a storm-threatened sky, there was no sound. She wondered why she could no longer hear the relentless roll-and-rock that seemed to play continuously in her son's head. Roll-and-rock? Rock-and-roll? Whatever, it was gone now. She was alone, for now. She swung her nylon bag lightly, in a gesture of defiance against doubt. Soon this strange escapade would be over. Soon she would have her son again—alive, and in her embrace.

WITH MY MOTHER ON HER WAY, I CAN rest again, feel the infinite light, and listen to the music. I hear the songs of my youth even now, even through the haze of this dream. I hear their beat, demanding my complete attention. I hear their lyrics, the messages that describe pain and a solution to pain at the same time. I hear truth running just above the melody and somewhere beneath the words, always discreet, but always present. I belong to the music and to the life that beats within it.

It's been a long journey through existence, a journey that started sometime before I could appreciate music—in fact, before hearing connected me to the physical world—and before I was aware of the struggles of men and women. It started before I knew anything of matter. My actual memories might have begun at the birth of my body, my initial attempts to breathe, and the sounds of my mother's anguished cries. From there came the eventful ride from infancy to manhood, from student to master. I have traveled from pure potential to the thrill of physical being to a road-weary ending. I have gone from endless nights of lovemaking to this quiet night, with death whispering within and around me. It's been a good life, a life of giving and receiving love without condition and beyond justification.

Love needs no justification; it is simply what we are. Men and women rarely allow themselves to feel the force of this. They know love only as a fallen symbol—a symbol meant to represent life, but one that has become corrupted by the many distortions of meaning. With the corruption of that one word, all symbols fall into confusion. Symbols grow into beliefs, and beliefs grow into petty tyrants that demand human suffering. All of this began with the fall of the first word: love.

There were many loves in my life, of course. There were always women eager to be touched, hungry to love and be loved. There were always women searching to see the truth of themselves in my eyes. In my life, I've loved them all. They had different faces, different names, but to me there was only one—only the fallen one, caught in a web of distortions and looking for a way back to truth. She seeks a path back to heaven even now, all the while believing the lies that keep her in hell.

Of course, she is all of us. She is Knowledge; and I can say now, without shame, that there was a time when she was Miguel. I had a good relationship with knowledge from the beginning. From my first breath, I was eager to learn the ways of sounds, symbols, and scribbled lines on paper. Like any healthy infant, I saw and heard everything. I felt in ways that adults around me had forgotten to feel. Sensation washed through me night and day, but clearly, sensation needed someone who could give testimony to its wonders. According to what I observed of the adult world around me, sensation needed a storyteller.

Feeling the flush of excitement that came with my first uttered word, and the thrill of seeing how it sent happiness racing through my parents and our friends, I was hooked. How quickly I became a devotee of words! How rapidly I used words to create a caricature of a little boy! Amazing, too, how words became the endless testimonial that is thought. In a very short time I grew exactly like those storytellers who populated my little-boy world. I happily collected assumptions and opinions, and the reward for my efforts was an incontestable identity. I knew myself well. Everyone else who knew me, knew me well, too—or so I believed.

I loved words and the universes that words created for me. I loved the power they gave me to convince other minds and change points of view. I loved the way words made it easy to romance girls and persuade knowledge-hungry boys like me. I loved the advantage that words gave me in school, both with peers and with teachers and then professors. I was always a good student. I was quick to memorize and quick to recall facts to mind. I was quick, that is, until I entered medical school. There, it seemed I had no advantage. No matter how hard I studied, how well I memorized, I could barely pass a test. My grades were poor, my temper was bad, and my self-confidence was plummeting. I wanted so much to follow in the footsteps of my brothers, but after my first semester in medical school, my prospects for a career as a doctor weren't looking good. Things got so bad that my physiology professor approached me privately, asking why my grades failed to reflect the intelligence and enthusiasm I showed in class. I had no good answer. I told him how hard I was trying and how much energy I was putting into memorizing the material. He stopped me there. "Don't memorize," he said. "Use your imagination."

This may have been the first time I heard words used in this way—to invite, rather than to convince. That professor was inviting me to break away from structure and to dream my life. He was giving me permission to *experience the truth,* not simply to *observe the facts.* My grades improved drastically after that—but, more important, the world as I knew it changed. This was the first of many steps away from knowledge, away from the compelling voice in my head. It was a small step, of course, because I was strongly bound to the laws of knowledge and, at that age, was knowledge's greatest champion. I believed it could cure every illness and solve every problem. It defined me. I was knowledge, in all its youthful expression and tireless aggression. Without the me that was born of words and ideas, I could not exist—or so I believed.

Watching my mother make her way to the horizon and to her destination, I'm at ease. Seeing the distant tree from my present refuge among the branches of the Tree of Life, I feel only love. That tree, mirroring mine, is the symbol of knowledge—only that—and symbols have no influence on me. *Now* they don't, but there was a time in my existence when I would have given anything to free myself of knowledge's hold. I would say its *power,* but knowledge represents a false power, born in those exhilarating moments of infancy when language is perceived as the only path to paradise. From that first seduction, there seems only one way forward. This is simple human destiny, of course. Out of infinite light we are brought into physical being, flung into dark perplexity, and challenged to find our way back. There's nothing that says we must burn with the same frequency of light that brought us here—but would it be so impossible? Bringing light to the obscurity caused by words is a determined choice, the path of a seeker.

My professor had asked me to dream the world from an academic point of view, but I soon discovered that dreaming is all we ever do. We imagine, and then we become. Those who are artists of the dream, whatever it may be, are artists of life. To dream means to construct reality, by whatever means available. A dog dreams the dream of a dog. A tree dreams itself in ways known only to the tree. It knows its body—every leaf and particle that makes it a universe. It knows the rejuvenating powers of sunlight, rain, and the nurturing soil. It perceives itself in relationship with all life, and it changes with the changing light, just as human bodies do. The human dream, on the other hand, adapts to changing knowledge. As human brains convert light into language, they learn to dream through words. We are gifted beyond our own understanding. Our words describe our reality. We are always dreaming, always redefining realities. In our sleeping hours, words are only the dim echoes of a waking dream, but the dreaming still continues. Like all creatures, we dream all the time. We dream an idea of who we are in relation to everything else. When other minds agree with us, we venture to call our dream *truth*. Depending on how we use knowledge, we can be victims or we can be responsible masters of our personal dream.

Just as I indulged knowledge so many years ago, there came a time when I had to refuse its authority. I had no cheering family then, and there was no community of humans to teach me how. I was alone, with only the ancient wisdom to comfort me. I was alone, as Sarita is now. Her journey to find me will begin in earnest in the world represented by that tree. Anyone can gather the pieces of an old dream, built by old knowledge. It takes a master to select the precious raw materials of a new and inspiring dream. This will be her challenge. She may fail or she may be victorious. Either way, Miguel will not be back. He is at home, here in the arms of eternity.

In his life as a man, he became aware of the truth of himself. Inch by inch, he slipped away from the temptations of knowledge. Ounce by ounce, he made his heart a weightless thing, emptying it of a thousand lies. The frequencies within him changed and intensified, until matter could not contain him. Revive the body, if you will, Mother. Gather the memories, bind them with faith, and let medical science do the rest. With eyes widened in excitement, see knowledge as if for the first time. Learn as you go. Be my heart in this quest, and grow lighter with every step. Do what you must. Try what you will . . . but Miguel will not be back.

Mother Sarita stood at the base of the second tree, feeling her heart battering against her chest as she gasped for breath. The tree had appeared so close, and yet the walk had seemed endless. Looking behind her, she could see the outline of Miguel's tree against the sky. It stood in the light. This one did not. Being a wise woman, she recognized obscurity for what it was. There was no evil here, only the absence of something. No, not absence: scarcity. Light was everywhere, and existed in all things, but light was not entirely welcome in this spot. The ethereal glow that flooded the surrounding landscape met resistance here. What had her son told her? He had said that she should put her trust in an impostor. She had no opinion about impostors. She had a job to do, and she would accept any help, in any form.

She took a deep, painful breath and felt her heart slow its pace. She had emptied the contents of her bag, but she felt the physical strain anyway. Strange, that this illusion should weigh so heavily on the physical senses. She was certain that, in the living room of her home, her heart was pounding in just this way. It might be that her sons were anxious for her, and that this trance state was frightening her grandchildren, but she could not stop now. She must keep going. She took another deep breath and tried to relax the muscles in her face, hoping that a calm expression might reassure her family as they watched her at home and wondered.

Seeing nothing in the branches of this tree but shadows and deceptions, she took a seat on a huge root that had broken above ground in one spot, arching like a cat that waits for a human touch. Just as she sat upon it, she sensed movement, deep among the branches of the tree. She remained still, pulling a handkerchief from her pocket and dabbing her face with slow precision. She sighed audibly and waited.

"Welcome."

The voice was silky and soft, but shocking all the same. It was both kind and cautious. It invited, and yet penetrated her thoughts. Its tone was sweet, but its message unyielding. With one word, it opened worlds. It was much like the voice of her son.

"Miguel?" she asked tentatively, her voice quivering. Was he in two places at the same time? What was this game of his, this dream of reflections? She worried that the ancestors might not approve, and she would need them before this excursion was over. Sarita stayed where she was, unsure where to look for the speaker, since the voice seemed to come from all places at once.

"You have made yourself comfortable," the voice declared.

"I am quite uncomfortable, as anyone could imagine," the old woman said, folding her damp handkerchief. "I doubt I could be *less* comfortable, but it is of no consequence, as I will be gone from here soon." From the corner of one eye she could see something slip smoothly from behind the trunk of the tree, not more than six feet away from where she sat.

"Oh?" said the voice with interest. "Where are you going?"

"I was told that *you* know better than I where I'm going." Sarita had an uneasy feeling that she was losing control over this trance. She had willed herself into her son's fevered dreams as a desperate resort and was now feeling the danger of it. Whatever risks she must confront, she knew she could reach him. She knew he would respond to her. She knew many things, but she did not know what she was facing at this moment. "Is it true that you know . . . well, that you know—" she faltered, unsure how to finish.

"I know everything," the voice said pleasantly. "Yes, I know *everything*."

Sarita was overcome by the feeling that this was no longer her son's dream; nor was it hers. This was an old, old dream, long repressed in human memory. This had the look of an ancient dream, one where a snake edged close and whispered softly. She could still see the familiar planet in the sky, brightly lit, with wispy dreams breathing in and out from its fiery heart . . . but there was little *here* that pulsed with life. The tree loomed beside her, but it seemed not to breathe. This was a dimmer sort of dream.

Sarita jammed the folded lace into her pocket, determined to make this vision hers to command. She would stand up and face what she had come to face. Her body obeyed, and she was on her feet in an instant, her expression grim and her heart pounding harder than ever. What she confronted was wholly unexpected. There in front of her, lurking in the ponderous shade of this tree, stood a beautiful young woman, clothed in a simple dress.

"Ah!" Sarita exclaimed, not disguising her relief. "Good. Since you know so much, perhaps you can tell me how to return my youngest son to the living."

"He is deceased?" the woman asked, seeming both surprised and sympathetic.

"He is not. He remains in that tree over there, dreaming of eternity." Sarita turned, pointing to life's symbol standing grandly on the distant horizon. "I will not let him die until . . . until he is finished."

She turned back to her new acquaintance, only to find that the young woman had moved swiftly and silently out of the shadows and now stared at the other tree with fascination. Her chest rose and fell in excitement, and her deep-red hair flowed behind her, as if caught in a sudden wind. This was not just any woman, Sarita realized with alarm. This was a magical creature, filled with power. She resembled the woman Sarita had once been but could barely remember—a sorceress, who held life in the palm of her hand and kept death serenely at her feet. Before Sarita understood just what she was seeing, the young woman had turned back to her and was staring directly into her eyes.

"Finished?" she asked sharply. "You say he has not yet finished?"

"What?" Sarita stammered in confusion. How could this creature help her recover her son? What could she know of him? "No," she replied, suppressing her bewilderment. "He has not finished. He has not completed his work."

"What work is that?"

Such nonsense! Sarita marveled at the creature's ignorance, but felt a growing satisfaction that she had recovered her advantage. Miguel must continue to travel, to commune, and to merge with Earth herself. This was obvious. He was a messenger. He was meant to do this and many other things. His dream was growing, expanding, and it must not end now.

"He has not yet finished his work with the Mother of us all," Sarita stated.

"She is no mother of mine," the woman said distractedly.

"He has not finished sharing his wisdom, giving generously—"

"Giving to whom? To you?"

"To the world! He has not finished being the messenger he was meant to—"

"He has not finished being your attentive son, you mean."

"He has not finished being . . . what he is!"

The vision moved noiselessly toward her, breathing cool breath onto the old woman's face.

"Is he not one hundred percent what he is?"

"Can you help me or not?" Sarita snapped, exasperated. "I will have him back with me . . . with the world."

The lovely creature took in a quick breath and leaned toward the old woman, inspecting her carefully. "You require my help?" was all she said.

"I desire your knowledge."

Another breath. This time the sound of it hissed under the flare of lightning in the dimming sky. Her eyes flashed red, and then the softest blue, as the woman laughed, her hair tossing in that strange wind of feeling that only she seemed to arouse.

"And to think," she hissed again, "looking at you, one might have suspected trouble! You are no trouble at all. You are a kindred, *vieja*. You are my likeness, my sister, and you are welcome here with me. If knowledge is what you desire, I will immerse you in it!"

"You may call me Mother Sarita, as I am your elder. Do you have a name?"

"I, too, am old. Older than you, Sara. . . . Sara," the creature pronounced carefully, enjoying the sound of it—an ancient name with sacred roots. She paused to study the old woman's face. "Sara," she whispered again. "Impressive name, and well deserved. For this occasion, I shall take a name that reflects me well."

Sarita waited, contemplating the list of things humanity had called this one through the millennia, sacred names and obscene names.

"What to call me?" the beautiful woman wondered aloud. "And in which delicious language? Your language?" Her face took on a look of worry, then amusement, then resolution. "Call me La Vida." She glanced quickly at the tree on the horizon, and a grin transformed her.

"Ah, yes," breathed the old woman. "Life." It seems the creature had ambitions beyond her scope.

"Or . . . perhaps not that. I think I prefer La Luz."

La Luz? This, too, seemed wishful thinking. There was little enough light in this corner. Sarita nodded agreeably. "Of course."

"No," the woman corrected herself. "La Verdad. Call me that."

"As you say," Sarita shook her head as she moved to retrieve her bag.

"Wait!" The vision spun in place, the hem of her dress stirring ash where she stood. "The name must be grand! Romantic! Call me La Diosa!"

Yes, of course, thought Sarita. Why not call yourself a god while you stand in your own proud world of delusions? She remembered being told once of a popular nightclub in Guadalajara by that name, where women shamed themselves dancing on the stage half-naked. The picture amused her.

"You make me dizzy," Sarita said, sighing. "La-this. La-that. La-la-la." Imagining the naked women in the strip club, she felt the impulse to play with this arrogant creature. "Could you not simply be Lala? It has flash." The redhead turned to stare at her. Sarita hesitated, fearing that she had caused offense. "That is to say, it speaks of both light and liveliness," she amended.

"I am La Diosa," the woman stated with finality, and then forced a smile. "As we are sisters in this cause, I suppose I could allow you to call me . . . some lively thing."

"Good. Then where do we start, Lala? Should I prepare myself?"

"Stay as you are, dear," she urged. "Let us call on memory, that prince of truth, to lay a path for us!"

"But memory—"

"I know everything," Lala interrupted. "Remember that. Doubt me, and we have nothing—nothing but light and motion and . . . and fragile buds on an unnamed tree."

Sarita tried to remember precisely what Miguel had said about memory, but could not. Before she had time to consider what might be wrong with buds on a tree, her companion had moved toward her, swiftly, soundlessly, and was gazing deep into her eyes again.

"The resurrection of a dream," she stated solemnly.

"The return of my son," Sarita corrected.

"This is felicitous," the woman murmured. "The solution lies within my realm of understanding." She held her piercing gaze. "You were clever to seek me out."

"Well, as it happens—" Sarita began, but Lala was still talking, still staring.

"Be sure you are respectful."

"Yes?"

"Be mindful of my unique skills, my ways, and my laws. Listen to me."

Listen, but don't believe, Sarita reminded herself.

"Listen, and obey," Lala added.

Sarita was resolved to remain in this remote terrain, regardless of the company. She must linger here until her son could be swayed. "Of course," she answered demurely. "How do we begin?"

The creature brightened at the question. "How. Yes." She smiled. "How, what, and why. There is no progress without these things." She moved away from Sarita, apparently thinking. Sarita watched her and waited.

"We begin with the first memory," Lala announced suddenly, "and move on from there." She glanced at the old woman. "You brought a shopping bag," she said. "You must have anticipated this."

Sarita looked at the bag, dumbfounded. Was it going to hold memories, then? Was this to be among her mysterious instructions? She wanted to laugh, but kept her silence.

"With enough memories, we have a dream—a talking picture show of all that is true about a man. I will guide you through the memorable scenes, through each crucial bit of knowledge, and in time we will have gathered all the pieces necessary to solve the puzzle called . . . Miguel." Lala's voice held on to the last syllable of his name as a bow glides along the strings of a violin, letting the sound fade slowly, melodically, into silence. *Miguel:* the word seemed to conjure images of something familiar, something sorely missed. The air around them stirred slightly, bringing with it a hint of warmth and sound.

"We have very little time, madam," Sarita said emphatically, breaking the spell of the moment.

"Time is my creation," was Lala's response. "We have as much of it as I say." With that, she took the old woman's hand and gently helped her back upon the giant root.

Sarita, her hand still in Lala's, thought she heard the faraway patter of rain, but the sky had not changed. Clouds still billowed and streamed; lightning still flashed in the distance, its force reverberating through her body, but no thunder followed. She felt the woman squeeze her hand. Lala stood very still above her, her gaze fixed somewhere in the distance.

She was looking, Sarita saw, toward the Tree of Life, and her face held a curious expression. It was one of fierce anger and deep longing. It was definitely both, although Sarita knew that such feelings did not exist together in the natural world. The old woman looked at the place she had last seen her son, and wondered if she should have respected his wishes this time and let him be. That was something she had rarely done, but now . . .

Lala suddenly let go of her hand. As Sarita looked back at her in surprise, the darkness fell over them both, pierced only by a single, soft light. It was candlelight.

Sarita was no longer sitting under a tree in a vast and desolate landscape. She sat on an ordinary wooden chair in the corner of a small room, watching a man and a woman making love by the light of a single candle set in a fruit jar.

M Y CONCEPTION WAS AN EVENT SARITA OFTEN talked about. My birth was a little unusual, and things became even stranger from then on.

I was born into an unusual family, one whose ancestry could be traced to the Eagle Knight lineage of ancient Aztec people. Those designated to be Eagle Knights were respected as wise men and advisors to their communities. Like our communities today, theirs were comprised of politicians, soldiers, farmers, and artisans. When I use the word "artisan" in relation to these ancient people, I also mean artists of life, or *toltecas,* as they are called in our storytelling. A reality based on words wisely spoken and beliefs playfully chosen was their form of art. Among the wisest, there were individuals who fought against the poisonous fear that so often comes from human thinking.

I didn't always know what I could become, or might become, while I lived in this human body. It is true that there were spiritual warriors existing even in my immediate family. At the time of my birth, my mother was already a known *curandera,* and to her healing practices she applied many of the sacred rituals that she'd learned from my grandfather, don Leonardo. His father, my great-grandfather, was called Eziquio. Although "trickster" was the word most often used in describing him, both with humor and with a little fear, don Eziquio was seen as a living legend by the adults around him. To children like me, he was the shadow, the specter, and the all-seeing eye. We weren't sure what mischief he had made in his life, or what sorcery he might still be capable of making, but we children spoke of him in cautious whispers, just in case. His name, like the names of all ancient shamans, was breathed with wonder and reverence.

Even as a young man, I couldn't have dreamed that words like "shaman" and "trickster" would someday apply to me. I wanted a respectable life as a medical professional, contributing what I could to the overall health of humanity. Never did I imagine becoming the one my mother had predicted I would become, or that I would be serving humanity in the way she had described.

Like all her stories about me, the story of my conception was always told in mythological terms. Most of her accounts sounded that way: the stories she told about our ancestors were as reverential as those told about saints and angels. I never really believed my own story the way she told it to everyone, but in time I became aware that she saw it as *her* story. My conception and birth, my immersion into the world-dream, and my return to her, to the ancient beliefs and practices—all of it was about her. And in very important ways, she was right. It was about her. Her stories were expressions of faith—faith in herself. Seeing how she lived her life with this kind of faith was the greatest of all lessons for me. It appeared that she gave credit to God the Father, and that she yielded to the will of the

Blessed Virgin. It appeared that she was constantly begging the saints to support her cause. It appeared that way because it was necessary to appear that way—but her power over people, and over the events in her own dream, came directly from the faith she had in herself. That faith gave life to her story. That faith gave life to the sick. And it was that faith which gave life to me.

◉ ◉ ◉

Sarita recognized the couple immediately. The woman in throes of ecstatic lovemaking was herself as a younger woman, before Miguel was born. This woman was naked and full of passion. This version of Sarita was mounted on her husband's lap, moaning with pleasure. Both their bodies shone with the sweat of sexual exertion. The old woman watched them, her eyes swimming with tears as she mouthed her husband's name and smiled in recollection.

"José Luis," she said, speaking out loud this time. "*Mi amor . . . mi cariño.*"

Standing near the bed was the mysterious Lala, much as she had appeared by the tree. She stood silently at the edges of the dim candlelight for a few moments, regarding the couple without emotion. Indeed, she scarcely showed interest in what she was seeing, so when she spoke to the old woman in the corner, she sounded much like a tour guide at the botanical gardens, pointing out common flora.

"She has a good body for a woman of fifty," Lala stated flatly.

"Forty-two," Sarita corrected. "And look at him! I had forgotten . . ." She sighed and looked away as the vision grew uncomfortable.

"He was a child," said the younger woman.

"He was well into his twenties," Sarita responded defensively.

"And when you married him?"

"He was well into his teens. *Far* into his teens."

"And you, a mature woman and mother of nine. Nine."

"He was so much in love . . ."

"He was in love with an idea, as we both know," Lala said, looking past the couple in the bed to meet Sarita's eyes. "He was spellbound. The poor boy had no chance."

"Ideas are everything," Sarita mused quietly.

"I'm glad you agree." A silky smile touched Lala's full lips. This excursion, so sudden and so suspect, seemed already to be going her way. Ordinarily, she was loath to participate in matters like this. The scene before her was an unsavory one—primal, and sticky with the promise of life—but it was a necessary introduction to a dream that she would one day own and control.

The young man making love suddenly cried out with pleasure, and as he did, his wife screamed, arching her back and stretching her arms to the ceiling. She screamed again.

"What?" shouted her husband. "What did I do?"

"The light! Did you see the light? It came out of nowhere and stabbed me in the belly! My body burns with it!" Sara, the younger apparition of Mother Sarita, brought her hands down and hugged the man tightly to her.

"*Mi amor,* there is nothing to fear from light," he whispered.

"God has touched me. There will be a child."

Laughing out loud, José Luis took his wife by the buttocks and threw her back onto the bed. "It doesn't take a celestial light to tell us that."

"Yes, there will be another, and he will be—"

"The thirteenth?" he guessed, mocking her. They had already added three sons to Sara's swarm of children.

"Yes, yes! He will be the thirteenth. Don't laugh—there is divine power at work here! Thirteen!" she emphasized importantly. "Stop laughing!"

Sarita listened to the couple talking, listened to the bed creaking softly, and remembered. "Yes, this was the beginning for Miguel," she said, almost to herself, "but there was so much that came before."

"We are here to visit the events of your son's life, not the life of the woman named Sara," the other woman responded dispassionately. Her face showed no expression as she watched the couple on the bed.

"We have no business visiting this!" Sarita snapped, rising to the single tiny window, open to the cool air. Outside, night cradled the world in its massive arms. A few random stars pricked the blackness, and the silence was broken by the yap of a dog—once, twice, and then no more. Sarita let herself feel the loneliness of silence. The love José Luis had offered her was bold, committed, and constant. She longed for the sounds of it again, the big sensations of it. She could recall how generously he had loved her, but she could not remember giving love in the same way. Too often she had repaid his devotion with condescension. He was a respectful husband and a helpmate, in her work and in the raising of their children, but what had she been to him?

Lala nodded her approval. "That's right, look away. There are things at work here that exist in defiance of knowledge, making this moment distasteful. We had to return to the beginning, and I suppose this is a beginning of sorts." She glanced at Sarita, delight flashing in her eyes. "But as for me," she said, "I prefer beginnings like *this* one!"

Sarita turned back to the little room and was amazed to see that it no longer contained a bed, a candle, and two lovers. It was now a kitchen, flooded with morning sunshine. And there she was, standing by a wood-burning stove as a younger woman again. She wasn't pregnant, which surprised her. When was this, then?

A radio played music, and she was singing as she prepared the day's meal. The screams and giggles of children could be heard from the tiny yard outside the kitchen door. The sound of traffic blared from the street, as the old Sarita watched the scene, her mouth open in wonder. A toddler played near his mother's feet, alternately sitting with his tiny paint-chipped soldiers and standing, balancing himself, and then taking a few steps closer to the stove. His mother yelled something to the other children through an open window, and then turned to smile with pride at the child, who was hardly more than a baby.

"My sweet boy," she crooned. "How clever you are! How strong and beautiful and clever!"

Encouraged by the tone of her words, the child took another step, then another. A boy of five or six ran into the house, knocking over a chair as he swiped a tortilla from the counter. "Hey!" he called as he kept running. "The monkey is walking again!" With that, a cheer rose from the yard. The toddler recognized the sound of it and beamed with excitement. These were the same wonderful sounds of laughter that rose up every time he stood, every time he fell, and every time he babbled incoherently. When his family laughed, he laughed. A lifetime of laughter wouldn't be enough for him. With that kind of reward in view, he steadied himself, lifted his tiny arms, and reached his mother with two more trembling steps. Once there, he clung to her strong legs with breathless satisfaction, burying his face in the folds of her skirt.

"He is a champion!" his mother shrieked, and a roar went up outside. She laughed, the boy laughed, and the universe rocked with pleasure.

"There, you see?" said his mother, caressing his little face. "Strong, beautiful, and clever. The whole world knows it!"

Watching the scene, Sarita spoke with fondness. "Yes, these are the days when Miguelito first began to walk. He was leaving infancy and starting his life as a child." She let out a long sigh. "Like his brothers, who took so much pleasure in tormenting him, he would develop a strong talent for trouble." She smiled as the memories rushed toward her and the light in the room began to flicker.

"Stop!" Lala said, interrupting her recollection. "My dear, this is not just the beginning of tedious boyhood that we are witnessing. Listen!"

They looked back at mother and child, as baby Miguel reached a small hand toward the stove, then pulled it back at the shocking recognition of heat. Sensing danger, his eyes widened in surprise.

"Ay! No!" he heard his mother shout. "No! No, no, no!"

Looking up at his mother, the boy repeated the sound. "No!" he mimicked with serious precision. "No!" The response from his mother was immediate and theatrical. Pulling him into her arms, she ran from the house, shouting to everyone that the little genius had spoken his first word.

"That! Did you hear?" shouted Lala with animation. "That is the beginning!"

"Of what?" asked Sarita. "Yes-no? Hot-cold? *Mamá-papá?* The beginning of words, you mean?"

"The Word," Lala stated, almost reverently. "See how it goes? One word leading to another, and another, until you build a universe of perception." Looking into the old woman's eyes, she said, "This moment is the beginning of knowledge and the universe it will create. This," she added wistfully, "is the moment of my birth."

Her birth? Sarita marveled to herself. The creature from the other tree yields to the laws of birth and death like the rest of us? In any language, it is a simple thing to recognize that a burning stove will cause pain. *No!* is essential to a baby's education. She regarded the other woman with interest, noting the pride in her expression. Who was she, to be proud of another woman's child?

"Never forget it," intoned Lala as she sat down at the table. "If you wish to retrieve your beloved son, follow the words."

"Nonsense!" boomed a voice from the doorway. The two women looked up, startled, and saw an old man hovering just outside, standing in full sunlight. He was not tall, but he held himself with dignity, lending him the look of a much taller man. His hair was a delicate white, but there was nothing else about him that seemed delicate. He was lean, sinewy, and quite handsome in a cream-colored suit that spoke of another time.

"*Papá!*" Sarita exclaimed.

"*Papá?*" the gentleman repeated in surprise. "How could I be father to this honorable *abuela*?" He tipped his hat graciously.

"Yes, it is true that I am now a grandmother and great-grandmother," Sarita said, moving toward him, "and that you are long since dead and buried! Still, this is a joyful reunion!" She hugged him and pulled him into the room.

"What world have I blundered into," he asked good-naturedly, "where my children are great-grandparents and withered memories have blossomed anew?"

Sarita was unable to answer. Seeing her confusion, he thought it best to take charge. He led Sarita to the wooden table and seated himself next to her.

"Who is it we must retrieve?"

"My youngest. You remember Miguel," said his daughter, placing a frail hand on his. "He is slipping away from us. He has suffered a heart attack, one that would seem fatal to any whose talents were less than ours."

"If this is true, it won't be words that bring him back. It

will be the irrefutable force of life." He glanced at her companion, now sitting regally at the head of the table. Assuming the person to be a gentleman like himself, he nodded deferentially before turning back to Sarita. Then, with a jolt, he looked again. No, Sarita's friend was nothing like him. In fact, she was a woman—strikingly beautiful, with eyes like burning coals. She smiled at him, and her eyes glowed brighter.

"This is Lala," his daughter said.

"La Diosa," corrected the woman. She could wield no power over the dead, she knew. *They* were beyond temptation, beyond her reach; but with this one, as with them all, there had once been a time. . . .

"Beautiful, as always, *señora,*" don Leonardo said with a bow. Then he felt a slow wave of realization. Could it be that his daughter was not fully aware of the nature of this affair? Until he was sure, he would play the game with sincerity. "Did you start at the beginning, ladies?" he asked.

"Well," shrugged Sarita. "It was a beginning of sorts. We began before Miguel was born—at his conception, in fact—but we found the scene to be unsavory."

"And unrevealing," added her companion.

"Show me!" the man said; and as he said it, the morning sunlight was extinguished.

Without warning, the three of them were standing by that earlier bed, in the dark little room with a tiny window where two lovers laughed and sighed.

"If you please," uttered Lala, retreating to a dark corner of the room. "I will not witness this vulgarity again."

"Don Leonardo," his daughter objected, her body heaving from the effort. "We've seen this before."

"Have you?" he said, smiling broadly. "Have you really seen this?"

The nude woman was sitting astride her husband, enjoying

· 47 ·

the pleasure of their union. Suddenly, she shouted in ecstasy as she had done before, her arms in the air and her head thrown back.

"There. Did you see that?" said the old gentleman.

"Yes," she said, turning away to look out the open window. "And I felt it. I remember the moment well."

"And?"

"And . . . I felt the burning light as Miguel came into being." A star flared in the blackness, and Sarita became distracted. "How amazing," she whispered, clutching the window frame as she leaned her face into the cool night air.

"Yes! Your body felt it happen. A message was delivered, and creation began in you. Again! Little Miguel won the race. He was one among the tens of millions of spermatozoa to try—and he succeeded!"

"Are you here to make tasteless jokes? I have neither the time—"

"We are witnessing the making of a soul!"

In spite of herself, Sarita looked back into the room. "Soul? The poor and sinful soul?"

"No such thing! A soul is the epoxy that binds a universe—a matter of basic physics," he proclaimed. "Here you see a universe being born out of the cataclysmic division of two cells!" He paused to take a breath, pleased with himself.

"A boy's body is beginning," she mused.

"A body that will grow into manhood. The soul will see to that."

"And what of God?"

"Yes," said Lala from the shadows. "Tell us what you know of God."

"It is all God," answered Leonardo with a glance into the shadows. He pointed dramatically to the bed. "Is this not God in action?"

"No, tell me," Lala said, over the moans of the two lovers. "Tell me of God."

"We have had our promenade already, *señora*. I have nothing more to say about God. I am witnessing God."

"Do we take this memory with us?" asked Sarita, impatient to go.

"No," Lala flared.

"Yes, indeed!" said the man. "Let it be the first of many such events—events that describe the life of my grandson!"

"And the next event?" pressed the old woman, grabbing her bag.

"I have an idea," said Leonardo with a glint of inspiration in his eyes. Over Lala's grumbles of protest, the room seemed to spin, turning bright and then dark with each revolution. The sound of a woman moaning continued, becoming sharper and more urgent as another room lit up, this one with fluorescent lamps and shiny metallic objects.

"A hospital?" muttered Sarita, leaning against a glistening wall. "I need to sit down again." As her father pulled up a metal stool for her, the moaning stopped. They looked at the scene in front of them. It was not the death of her son she was witnessing now, but the moment of his arrival.

"Why the silence?" asked Leonardo. "Has he emerged? Has he been born?"

"He has," said Sarita, remembering. The conversation in the room had stopped with the mother's last push. All that remained were a few anxious whispers, as a nurse fussed over the newborn, coaching him to take a breath. The doctor busied himself with Sara, who lay still on the bed, pale as death and too exhausted to listen for the sounds of her infant son.

"They thought we would both perish that morning," the old woman recalled.

There was the feel of tragedy in the room. The needs of the

mother had become urgent, so the nurse holding the baby was called to help. She laid his lifeless little body on a metal table, an offering to fate.

"I smell fear," Lala remarked. "Lots of it." She moved from the far wall and stood at the center of the room, her elegant nose in the air. "Yes, fear . . . mixed with blood." She backed away, repelled.

"Not to your taste?" the old man goaded.

Lala ignored him and cast her eyes around the room disapprovingly. Blood was everywhere. It covered the bedsheets, the inert mother, the anxious doctor. It had splashed onto the white-tiled floor and smeared the metal surface where the body of the child lay, cold and silent. It smelled, yes. It smelled of copper mines and manure. It smelled of fertile things—secret, undiscovered things. It smelled of life.

"Not to my taste, no," she conceded. "I prefer my world of named things over the world of oozing, writhing things."

"Yours is no world at all, my dear."

"It is exactly the same world you once occupied—but without the detestable mess."

They regarded each other suspiciously, and the silence in the room grew heavy.

"Father," Sarita exclaimed. "I can't bear the horror again! He doesn't breathe!"

"Wait, *hija*," said the old man. "Here it comes . . ."

Don Leonardo stretched his right hand toward the lifeless infant, palm open, and there was movement, then the unmistakable signs of struggle, as the baby's frail lungs billowed and contracted, sucking in air. With the next gasp came shock, then sound, as the boy announced his existence with a vigorous scream. Half-conscious, his mother cried out and lunged toward the sound, almost tumbling from the bed. The nurse dropped her tray of soiled towels and shrieked in alarm.

"Ah, no. He wasn't going so easily!" said Leonardo with a laugh. "Nor will he today, my daughter."

"This is a waste of precious time," Lala said, raising her voice in a show of authority. "Am I in charge of this expedition or not?"

"Please, yes," Sarita replied. She collected her nylon bag and rose to meet the woman in the center of the busy room, where doctors and nurses celebrated the miraculous resurrection of mother and child. Sarita had a mission, and very little time. If this mysterious woman had the answers, she must be obeyed.

She nodded to her father, and the three spectral guests walked out of the room, the older woman taking the lead. Leonardo gazed back at the operating room, marveling at the chaotic wonder of it. He thought he saw someone familiar standing by the wall, but before he could get another look, he was pushed abruptly from the room. Lala followed close behind him. She, too, hesitated by the door, and turned.

Miguel Ruiz was clearly visible, standing in the light of an operating lamp. He was a grown man, radiant though dressed in a hospital gown, just as Sarita had last seen him—a man whose recent heart attack had hurled him out of the human play and into a world between two worlds. His gown showed flecks of blood, the blood of his own birth. He bore the stains of humanness that Lala so abhorred. She suspected that he had come here to feel the invitation. He had come to remind himself of the thrill—of the fearless daring a newborn feels as he launches headfirst into the human dream, taking a sharp breath and crying out in delirious exultation. He had come to watch, and to imagine.

Miguel and Lala looked at each other wordlessly. Each recognized the other, as any person might recognize himself in the mirror; but there was something more to the way they regarded one another. In Miguel's eyes shone the full expression of love, without fear or doubt. In her eyes lay suspicion and the expectation of loss.

It was hard to say in that moment whether either of them could sense an opportunity for union, for laughter, or for the sweet submission to desire. It was hard to imagine how many directions this journey might take. Lala seemed as capricious as any woman, and just as eager to steer the events of the dream her way. Hearing Sarita's call, she offered Miguel only the slightest of smiles, and then she was gone. Miguel stayed where he was, watching the hospital doors swing shut behind her, and let his imagination carry him gently into another dream . . . a dream of times forgotten and feelings exchanged.

I T GIVES ME COMFORT TO WATCH SARITA NOW . . .
with her esteemed father again, and so present within my
memory. When I was a small child, my mother was the
only woman I really knew. I had older sisters, but they were
already married and remote from my everyday life. I adored
my mother, and respected her above all other beings. She was
beautiful, wise, and pure. She was the Virgin, as every woman
was in my young imagination; and as I matured, I would hold
all girls to the same standard.

Growing up, I saw how my older brothers acted with their
girlfriends, and I envied their cool, their gift with the opposite
sex. I was amazed by their apparent confidence—it seemed
they had special talents and rare insights into the minds of
women, and I hardly hoped to achieve their success in romance.
Well, hope is a trickster. It feeds illusions to hungry hearts,
much as my great-grandfather Eziquio did. It seduces the mind

with promises it cannot keep. As it happened, though, it wasn't hope that made me a success with women; it was action.

My romantic life began at six years old, when I spontaneously asked a pretty classmate to be my girlfriend. Her immediate response was to laugh in my face. A few days later, when she reconsidered the offer, it was my turn to laugh. I rejected her. Yes, already I had learned to reciprocate the pain, a typical stratagem for emotional survival.

It seemed like a lifetime before I had the courage to try my luck with the opposite sex again. Before I turned twelve, however, my brothers had experienced enough pain of their own to sympathize with me. Jaime, the one closest to me in age, insisted that I try again. He gave me a motivational talk one morning, explaining how I was sure to get *one* girl if I was brave enough to ask ten or twelve. Whoever accepted me wouldn't be the most desirable, of course, but my confidence would be restored. So, with borrowed courage, I asked a shy little friend at school to be my girlfriend. She said yes immediately. I was stunned. On my way home from school, still delirious with excitement, I asked another girl. She also said yes. By the end of the week, I had eight girlfriends and no idea what to do with any of them. They all seemed happy with the arrangement, as was I. My confidence, only just discovered, soon turned to expertise. Even Jaime was amazed, though the rest of my brothers were merely amused. They still teased me, but for the first time, their jokes carried some male pride and approval.

I was a little guy, but slowly I became a rock star in the arena of scared boys and giggling girls—all eager for romantic stories to tell. It wasn't long before I became a favorite among the older girls. Sweet words ran like guava juice off my lips, making them all laugh and blush and warm to my little boy kisses. I was cute and funny, and they told themselves I was too young to be dangerous. Sex is a simple enough thing when fear doesn't intrude on the moment, and with one blissful moment of success, innocence was happily lost for me. I would never again be hungry for love. After a short lifetime of poverty, it seemed that I was on my way to becoming a sexual billionaire.

I say all this to make a point about seduction. Seduction is a skill conspicuous to all living things, and one that is vital to life. Just as it works in the natural world, so does it work in the universe of thought. An idea spoken fearlessly causes a contagion of agreement. An invitation spoken sweetly erases any sense of danger. Suggestion provokes imagination, and imagination builds reality. When we can see these things clearly, we can also see beyond words and suggestions, to the messenger. Any messenger uses knowledge to gain access to a dream. "What do you know? I know it, too," is one way to start. Or "What do you like? I like it, too." Once invited in, the messenger can begin to change the shape of that dream. It is an unusual messenger who uses seductions of the mind to benefit another human being. It is an iconic messenger who applies this skill to benefit humanity as a whole.

The creature my mother met on her visit to the Tree of Knowledge is and always has been a skilled messenger. She has been moving and shifting the human story for as long as that story has been told. I refer to her as a woman, not for the reason the world does, with its peculiar distrust of feminine insight, but because I recognized early on that, like most men, I was born to love and cherish women. Once I tasted love, I never stopped wanting it. As a young man, I had a similar infatuation with knowledge. Just like a clever woman, a woman of remarkable power, knowledge captivated me. I suppose I was spellbound and obsessed for many years, but once I saw knowledge for what she was, I used all my talent to break the spell. I felt a desire to redeem knowledge, to guide her into awareness and to live with her in peace. I used the talent that came naturally to me—my talent for romance. By seeing knowledge as a woman who wanted above all to be known and to be heard, I could begin to listen, and to take the fury from her. By recognizing her need to be loved, touched, and tasted, I could transform her.

After I became a shaman, I finally saw that this was the revelation that my grandfather, don Leonardo, most wanted me to have. I finally understood his words, always understated and free of pretense. They weren't like other wise words, laced with charming guile. Such guile is the character of knowledge, making it a clever messenger, but not a messenger for truth. My grandfather was a man who had heard the voice of knowledge in his own head, and then silenced it. In that silence, life finally made itself known to him. In that silence, he met his own authenticity. The wisdom he was able to share with me was wisdom he had achieved by seducing the temptress. Knowledge is the thing that moves men and women to think and behave as they do. Its authority begins with our initial attempts to speak; then, as we master language, it evolves into thought. It becomes the voice we listen to most, our most trusted informant. Knowledge gains power with every belief we embrace, regardless of that belief's impact on the human being.

We master death when we finally know ourselves as life; when we can see from the perspective of life, not just knowledge. Each of us is the main character our story, and the main character is afraid of not knowing and not being known. Death represents the greatest threat to knowing and has therefore assumed a terrifying significance in the human dream. Death to an individual means the end to the physical body and the conclusion of thought. Death doesn't mean the end of life as a whole, however, nor does it mean the end of humanity.

When knowledge serves our fears, it can make the sensible seem satanic, the satanic sensible. And yet knowledge, the single greatest devilry for humanity, can also be its savior. It's up to each of us to recognize knowledge as the voice in our own head—the voice we have come to trust and obey. It's up to each of us to modify that voice and to reform the tyrant. For, in the process of mastering knowledge, we have become knowledge. We have become the tyrant, the tempter, prompting fear with every opportunity. By redeeming knowledge, we redeem ourselves.

I have the feeling I should be elsewhere," said don Leonardo as he paced the little schoolroom. He was walking between rows of desks, absently glancing at first graders as they scribbled symbols on paper.

"I need you here with me," Sarita reminded him quietly. She was sitting at one of the desks, her body jammed between bench and tabletop as she watched the teacher in action.

"Shush. Listen to the teacher," chided Lala, her red hair swept high on her head in the fashion of the day. "What she says is important."

All the same, Lala was looking at the young teacher with disapproval. "Why is her appearance so shabby?" she asked her fellow itinerants. "Teaching is the single most important job there is, yet she brings no style to it! She wears no heels, no rouge. Learning should arouse us, should it not?"

"Not the kind of arousal I'm familiar with," said the old man, straightening his tie as he checked out the teacher, who was wearing a cardigan and simple skirt. "If I may imagine her without the clothes—well, then . . ."

"*Papá,* the children!"

"Hardly more than a beast," the other woman murmured, her voice silky and contemptuous.

"That's enough!" snapped Sarita. "Because we are among first graders, that does not mean we should act like children!" She scanned the room. "Where is Miguel, the six-year-old?"

"There, by the window, daydreaming," replied Lala. "But he is not the point. Listen to the music of knowledge, strumming a song of power and possibility."

"Welcome to your first day of school," sang the teacher, her face alight. "My name is Señorita Trujillo, and I know you are nervous. Some of you may be afraid, and others excited, but you are all here—like your parents and brothers and sisters before you—to learn how to be people in this great society."

"To be people . . . ?" whispered Leonardo.

"Because they are hardly more than beasts," repeated the redhead.

"And I am expecting each of you to work hard," Señorita Trujillo continued. "If you work very hard, you will reach perfection, and perfection is what we all desire."

Leonardo stood in place, pivoting on his feet to inspect the twenty little boys and girls. "And how are they not perfect?" he asked. "How," he said, his hand gesturing to encompass the perfect heads of the children, "can these angels be considered imperfect?"

"They have learned nothing yet!" argued Lala. "They barely know how to think, how to judge. They are slow to make assumptions and quick to ignore sacred beliefs."

"Are you now an advocate for the church?"

"I have always been a friend to religion," she said haughtily, "and I fully support God's rigorous judgment."

"Amen," said Sarita, crossing herself. She considered it a good habit to agree with anything stated piously. As she kissed her own thumb, she noticed the disappearance of morning sunlight, and found herself seated on a pew within a chapel murky with smoke. "Are we now in church?" she asked, disoriented.

"Ah!" cried don Leonardo. "You go too far, *señora!*"

Lala glared at him, her eyes burning. Sarita looked from one to the other in surprise. "What are you doing, you two?" Her eyes wandered to the pews. "This place has nothing to do with my son, or with his memories."

"Indeed, it does," answered Lala, happy to draw attention away from don Leonardo. Just then, they saw a priest walking toward the front of the chapel. He walked past them, unseeing, and continued to the first pew, where Sara, the young mother and wife, was sitting quietly. Nearby, her four boys moved from bench to bench in a silent game of tag.

"Ah!" exclaimed Sarita. "There am I, with my boys!"

"Remember this?" asked Lala. "When the good priest told you your thirteenth child would make a difference in the world?"

"Yes. Yes, I do. He said Miguel would be an important messenger."

"Was that before or after he judged him a sinner?" asked Leonardo.

The redhead ignored him, leaning closer to Sarita. "And what is a messenger, but the servant of knowledge?"

"An authentic human," said don Leonardo flatly. "Such is a great messenger."

The woman gave him another angry look, but Sarita was talking, apparently reminiscing, and their attention was drawn back to her.

"I remember when he grew old enough to feel his own power," she was saying, "a power that had already become evident to me. He was a boy of ten then, older than he is here. I had been talking to him one day about greed and selfishness, and how we hurt ourselves when we disrespect others. When I was finished talking, he looked at me with all the serious- ness of a worried mouse. 'Do you think me selfish, *Mamá*?' he asked. What could I do but laugh? 'Yes, sweet one,' I teased. 'You are as selfish as your mother is supremely generous.' It was a hasty joke, I admit, but he paid no attention. He was lost in thought, sorting truth from subtle lie."

Don Leonardo sat beside her and took her hand, hoping to encourage the memories.

"He smiled at me," Sarita went on, "and it was a gracious smile, a smile that was conscious and careful. I had the im- pulse to touch Miguel protectively, but this particular smile told me not to. It told me that he was now old enough to grasp the truth with his own small hands. If he was selfish, he would find the remedy. It reminded me of something I had always known, even without the counsel of the priest: that humanity would one day long for his words, for the touch of his eyes, his hands, and his irresistible smile. The smile he gave me on that particular day told me that I was becoming a stranger to him. The tender bond that held us was weakening."

"It was strong then, my girl," her father assured her, "as it is now."

Lala turned away impatiently, wishing the priest would speak to the young Sara, that he would say something about the excellence of minds, not this nonsense about emotional bonds. What was the purpose of this particular memory, ex- cept to remind them of the power of words?

"The point is—" she began, addressing father and daughter again.

"She sees the point, *señora*," stated Leonardo.

"My love will bring him back," Sarita said softly. "Our bond, which cannot be undone, must now be respected. His legacy must be—"

Lala grabbed her chance. "His legacy exists in the minds of everyone he has touched! He is memory. He is thought, and his words echo through the ages."

Don Leonardo heard Lala's voice ascend into the high dome of the chapel, and he refrained from comment. He squeezed Sarita's hand supportively before rising to his feet. It was Sarita's challenge, not his, to resist the persuasions of that voice.

"Indeed," the old man mumbled loud enough to be heard, "there is definitely someplace I'm supposed to be. Good day to you, ladies," he said, smiling, "until the next time." He tipped his hat and marched up the aisle, toward the light that beckoned beyond the chapel doors.

"*Papá!*"

And he was gone.

Sarita looked back, confused. "Now what? I need his help."

"You need *me,*" said Lala. "We already established that. We will follow the same path of esoteric knowledge your son followed. We will follow his most remarkable thoughts, and in that way—"

"Thought is knowledge, Miguel would say," Sarita interrupted. "Memory is knowledge, he would say." She looked at the other woman, so beautiful and so certain, illuminated by the light of a hundred prayer candles. "Religion is knowledge, my son would say."

"And see how wonderful that is!" replied her companion, indicating the chapel, now filled with kneeling worshippers, many of whom were weeping silently.

"None of it is the truth, my son would say."

"Your son will be back with you soon enough to speak for himself. Come, old one, and we will find the precise day his words showed him to be a leader."

Sarita stood up, crossed herself again, and followed the beautiful woman into a soft cloud of incense.

D on Leonardo was now exactly where he was supposed to be. He was watching one of his grandsons walk up the path to his house. He was a handsome boy of twenty—not the youngest of his grandchildren, but Sara's youngest, and that was enough to make Leonardo smile. He loved his daughter and had always seen exceptional things in her. She could have succumbed to the monotonies of human life and old habits, but she had maintained her singular authority. She would soon be recognized as a woman of power by her family and her community.

This boy, too, was not like the others. Leonardo knew this, but could not yet say why. All Sara's sons were quick, bright, and full of ambition to succeed. He remembered the morning of Miguel's birth and wondered if the answer to his uniqueness lay there, but too much importance was given to family stories like that—too much was made of auspicious signs and star alignments. The answer was in each present moment, like this one. At this moment, Leonardo was watching his grandson arrive and he perceived everything. He noticed the boy's stride, the confident way he held himself, the light in his eyes. Yes, the eyes told much. And the smile. What a smile! It seemed to invite the world to play. Come into my dream, it said, and prepare to have fun!

This boy was different, certainly. It was time for Leonardo to see if that difference could be put into exceptional practice. It was time to take this boy out of the sleepy warmth of his convictions and into the icy air of awareness.

Don Leonardo stepped off the porch and opened his arms to embrace Miguel, the last of Sarita's thirteen children.

I remember that beautiful autumn afternoon during my first year at the university, when I rushed to my grandfather's home to pay him a visit. My heart was filled with love that day, and my head filled with new ideas. He was in his nineties then, the leading elder in our family. He was respected by everyone who knew him, and I felt tremendous pride in having him as a grandfather. I also felt great pride in being able to go to him, to talk to him, and to share my knowledge. I wanted him to think me wise for my age. I wanted him to be impressed by my penetrating intellect.

I have learned since then not to show gratitude to a great master by offering him knowledge. Offer anything else, but not that. If you have even a little sense, you will offer him your silence. He is a master for the simple reason that knowledge is not a distraction to him. You are the student because it is, to you! In my life as a shaman, I couldn't count how many times, and in how many ways, my students tried to impress me, just as I tried to impress don Leonardo that autumn day. So many apprentices with a potential for deep wisdom would fall short, choosing instead to serenade me with facts, opinions, and philosophical references. "How can I amaze you?" they might as well have said. "What do I know that you haven't thought of? Look at me! Listen to me! Let me teach you something!" Do we sit at the feet of a master to celebrate our own importance, or do we sit with a master to listen and to learn?

Well, I didn't visit don Leonardo to listen that day. I started talking from the moment we sat down in the yard together, and it seemed I couldn't stop. I told him about all the political activities at my school, about everything I had learned of governments and politics, of injustice and human suffering. I spoke with righteous anger and moral indignation. I spoke against humanity for all its many evils . . . and it was just about then that my grandfather's smile turned to quiet laughter. There was nothing funny about what I was saying, so it seemed clear he was mocking me. Mocking me! Had he grown so old that he could no longer see the brilliance of my logic? Couldn't he see how insightful I'd become? I stopped talking, feeling the shame build in me.

"Miguel," he said gently, with a sweet smile on his face, "all the things you've learned in school, and everything you think you understand about life, comes from knowledge. It isn't *truth*."

Didn't he realize that I was a man now? He was speaking to me as if I were a child. I felt heat in my face as his words began to anger me.

"Don't take offense, my child," he went on. "This is the mistake everyone makes. People put their faith in opinions and rumors—and out of this, they construct a world, believing that their constructed world is the real world. They don't know whether what they believe is true. They don't even know whether what they believe *about themselves* is true. Do you know what is true, or what you are?"

"Yes, I know what I am!" I insisted. "How could I not know myself? I've been with myself since birth!"

"*M'ijo,* you don't know what you are," he said calmly, "but you know what you're not. You've been practicing what you're not for so long, you believe it. You believe in an image of you, an image based on many things that aren't true."

I didn't know what to say next. I had expected praise, or at least an argument against my point of view. I would have been happy to participate in an intellectual boxing match with my grandfather. In my opinion, I had enough information to debate the master, and to win. Instead, what he gave me was a knockout punch to the self. Everything I thought about Miguel, my grandfather disqualified in a few hard sentences. Everything I knew about the world was now in doubt. Doubt!

It's hard to overstate the importance of doubt when we're bringing down the intellectual house we've built. We learn words, we believe in their meaning, and we practice those beliefs until our little house is solid and strong. Doubt is the tremor that brings it down, when it's time. Doubt can cause a citadel of beliefs to crumble; and that kind of tremor is necessary if we want to see beyond our private illusions. An earthquake is necessary. I looked at my grandfather, and he smiled back at me, as if we had just shared a happy secret. Did he even notice that my self-esteem had been shattered?

"I know what I am, and I know about . . . things," I stumbled. I was feeling defiant, as if defiance would save me from my embarrassment. "I know about the world I live in, and I know that good must always fight against evil."

"Ah!" he said, with a flush of excitement. "Good versus evil, yes! The age-old human conflict! Do you see this conflict in the rest of the universe? Do you see good and evil wrestling within the forests and the orchards? Are trees anxious about the evils of the world? Are animals? Fish? Birds? Are any of Earth's creatures consumed with worry over matters of good and evil?"

"Of course not."

"Of course not? Then where does this conflict exist?"

Was this a trick? Was he determined to make a fool of me? "In the human species," I said warily.

"In the human mind!"

"Well, yes . . . and there's nothing more noble than the minds of men," I added pretentiously. "If animals—"

"If animals could think, they'd be as worried about evil as we are? I hope not, for their sakes!"

We both laughed, and for moments after we were silent. "Miguel," he said, when he felt my defenses weakening, "the conflict you speak of exists in the human mind, and it is not actually a conflict between good and evil; it is a conflict between truth and lies. When we believe in truth, we feel good and our life is good. When we believe in things that are not true, things that encourage fear and hatred in us, the result is fanaticism. The result is what people recognize as evil— evil words, evil intentions, evil actions. All the violence and suffering in the world is a direct result of the many lies we tell ourselves."

I suddenly remembered the words of a great philosopher: *Men are tormented by their opinions of things, not by the things themselves.* I couldn't remember where I had read that quote, or who had said it. A German, perhaps. No, a Frenchman.

"Miguel, stop," don Leonardo said sternly, bringing me back from my fixation. "Stop, please," he said, patiently this time. "Great thoughts should be applied, not catalogued. The privilege of knowledge is to serve the message of life. Knowledge itself is no message at all. Left in charge, it will drive us mad."

I could sense that he was right. After a moment, I told him so, and he leaned back in his lawn chair and looked at me for a long time, considering. I thought the conversation was over, and that by agreeing with him I would be released. I could grab an *empanada* from the kitchen, say goodbye, and ride back to the city, where people appreciated me for my intelligence and wit.

"Miguel," he said, his expression so serious that I knew I wasn't going anywhere. "I see you're trying hard to impress me, to prove you're good enough for me, and I understand. You need to do that because you're not yet good enough for yourself."

Tears rushed to my eyes. I saw right away that my determined efforts to appear confident were a waste of time. All my opinions and assertions were hiding the fear that I wasn't wise enough or smart enough. Don Leonardo could see more than I could see, and knew more about myself than I was willing to discover. I looked away from him, unable to handle the truth in his penetrating gaze. I looked away, yes—but I stayed where I was. I stayed with him to listen.

He told me much that afternoon, and it has taken me a lifetime to digest our conversation. What each of us wants above all is the truth, and it cannot be told in words. Like everyone, like everything, truth is a mystery posing as an answer. The letters I learned in school point to revelations that point back to mystery again. Truth existed before words, before humanity, and before this known universe. Truth will always exist, and language was created to be its servant. Words are the tools of our art, helping us to paint images of truth on a mental canvas. What kind of artists are we? What kind of artists do we want to be, and are we willing to give up the nonsensical things we believe to become those artists?

My grandfather told me that my greatest power was faith. It was up to me to direct that power wisely. The world was full of people eager to put their faith in an idea, an opinion, the opinions of other people. He urged me not to invest my faith in knowledge, but to invest it in myself. Though I didn't realize it then, our conversation that afternoon set me on a path I would never abandon. From then on, I wanted to make sense of things. I wanted to understand myself and find out how it was that I had begun to believe in lies. It was my nature to seek answers. It is everyone's nature to find the truth, and we will eagerly look for it anywhere, everywhere—except in us.

I wanted only the truth after that day, and all I had to guide me in the beginning were memories—memories based on random images and stories, leading to more distortions. But that was only the beginning. How quickly things would change for me! How generous truth is when we are willing to feel it, accept it, and be grateful.

Sarita, my lionhearted mother, is taking a similar path on this long, dream-fueled night, guided by the same memories . . . while the voice of knowledge whispers earnestly in her ear. For her troubles, she will bring home a pretender—the flesh-and-blood likeness of her youngest son, who has already found the truth, and has gleefully dissolved into its wonders.

S ARITA WAS TIRED. SHE HAD BEEN LISTENING TO
the speeches of a dozen or more student activists on the
university campus. Miguel had been the second to last
to speak, and he was something to see, rallying the crowd to
this cause and that one; but she was tired now, and unsure
how all this would help her get him back. She removed one of
her slippers and gently rubbed a swollen foot. It would be a
long night, she knew, but it could not last forever. Her grandchildren would be asleep by now, their parents still drum-
ming by candlelight, still watching Mother Sarita as she held
the trance and continued her peculiar journey. This was hard
for them, too.

"I know he was a good speaker in college, Lala" she commented to the woman guiding this expedition, "but this is not
such a special day in his life . . . nor would my son count it as
memorable."

Sarita fidgeted, feeling uncomfortable within these surroundings, as she was reminded of things she had long forgotten. Escaping the night massacre of Tlatelolco—*that* was memorable, she thought to herself. Miguel and his brothers, students at the Autónoma Nacional University in Mexico City, had traveled home that week and so, thankfully, were not in the Tlatelolco neighborhood when the military opened fire on thousands of students and bystanders during a peaceful rally against government policies. The killing had continued into the night, ending in the tragic loss of many of her sons' close friends and professors. Yes, it was important to remember the young and vital ones who had been killed, whose promise would never be fulfilled; and it was important to be grateful for the lives of those who had avoided the massacre's horror. That was not the only time death had spurned her youngest son. No, he and death would face each other and depart as cautious friends many more times.

"Indeed, he was so young," Lala agreed, "but you see how persuasive he could be, even in his first year at medical school. He had a way with the spoken word. He had charisma. He brought his fellow students together, as we see. With such a forceful personality, he could have influenced a nation."

Sarita nodded, remembering how intensely her son had been courted by government officials in those days. His brother Carlos had advised him on the dangers of politics, and Miguel had been quick to understand how recruitment into that kind of life would compromise his personal freedom.

"I must find don Leonardo again," the old woman sighed, massaging the other foot. "He will know what is important to this quest."

"Men know about men, I suppose," the redhead muttered. "There's a good chance he's observing couples in bed."

"Is it time for that again?" Sarita exclaimed. It seemed that young men were unduly proud of their lovemaking, as if they thought they had invented the thing. She pictured Miguel as he'd been then, so young and so amorous. She thought of Maria, his wife, and their beautiful sons. Of course, sex came with great rewards—physical joy and the pleasures of parenthood. Nothing touches us more than marriage, more than birth . . . more than death.

Sarita lifted her head, slipper in hand. "Death," she said, turning pale. Looking away from the park, from the people, she saw something that had escaped her notice until then. In the distance, a young man was driving a junker of a car, weaving slowly through the crowd of students as if looking for someone.

"Memín," she whispered, her mind reeling with the memory of another son . . . and then she fainted.

⬤ ⬤ ⬤

Sarita," Miguel called softly, "*Madre,* are you there? Sarita?" From the depths of a dream, Sarita became aware of his presence. With eyes closed and a mind spinning in and out of worlds, she gave him silent assurance. She imagined him sitting in his tree with Earth blazing behind him; pictured him laughing at her as the madness continued. She could not bring him back against his will, nor could she stop trying. She had invested too much, and involved too many. She submitted to the crushing pain of a mother on the verge of losing another precious child. Miguel was near her, watching her, she knew. He was there and not there, just as she was. She could feel his closeness, his attention . . . but oh, how she wished to hold him again! She moved her lips, still not speaking, and yet somehow words were shaped, and they were heard.

"I am here, child," she whispered into the unknown. "I am with you, in you; and my intentions will not falter. Old as I may be, I still have strength. Frail as I am, I will conquer your resistance. Brave as you are, I will win."

Sarita felt an overpowering yearning, wishing for a glimpse of her son's face, the touch of his hand on hers. She felt his closeness then, as he seemed to respond to her wishes, and was comforted.

It hadn't always been like this between them, she thought, as she slipped farther into a dream state. There had been a time when the only thing the two of them could not tolerate was being apart. It had seemed a never-ending, enchanted time, one that had begun as soon as mother and son first recognized themselves in each other's eyes. From their earliest moments together, they were bonded by a force greater than love. Greater than love, yes. Love was a word corrupted by misuse and self-ish wants. It was a glorious gift sullied by conditions. Over time, the symbol of love strengthened its hold on the human heart like the grip of a lioness at the kill. It was true that their bond was greater than love, and far greater than the terror that has sometimes run like a jackal in the wake of love.

From the moment of her son's arrival, she had sung to him, and from that moment on they were as one. As Sarita now struggled to hold the connection between them, she remembered how the infant boy had lain naked in her embrace, wearing the bloody residue of his journey from the womb. His face was pressed against her damp breast and his tongue tapped at her nipple as he relaxed into his mother's scent and breathed to the rhythm of her heart. The sensation washed her in comfort. She submitted to the primal silence and marveled at his guiltless eyes. With her fingertips she traced the curve of his tiny face and the soft bend of his arms and legs. She caressed his smooth, amphibious flesh and wondered at the fragile warmth of him.

"Yes," she whispered aloud as she dreamed, "I cried happy tears to finally look upon the child I had conjured with a wish . . . and had hidden within me like a secret. You, my jewel, had arrived, and with your coming, all pain and worry passed from me. From that moment, we were joyful in each other's arms and never doubted that the joy would last a lifetime."

Doubt came, of course. It came later, and it has come many times over the years, as the bond that had once been so strong began to tear. It came the day Memín was killed. He was the youngest child from her first marriage. He was her treasure, and the symbol of heroism to his little brothers. That painful day led to many painful days, and by the end of it, she and her youngest child were changed forever. By the end of it, Miguel had begun to see humanity as it truly was.

◉ ◉ ◉

Qué pues! What have you done to my daughter?" don Leonardo demanded. The university campus was gone. Sarita was lying on the grass within a cemetery park, her bag clutched to her chest and a bare foot exposed to the sun. She was barely conscious, hearing noise but unable to derive meaning from it.

As she lingered on the edges of a dream, cars pulled up to the curb close by. People gathered near an elegant elm tree, all of them dressed in black. They exchanged quiet greetings and a few tears as they prepared to bury a loved one.

Lala, apparently unaware of the scene around her, knelt beside Sarita, stroking her gray hair and clutching her hand.

"I did nothing!" she barked, her voice strained with worry. Lala was feeling an odd sort of fear, suspecting that Sarita had become too exhausted to pursue her cause. She could not let that happen. Miguel must not be allowed to die. His existence was important to all of them, but few knew how important he was to Lala.

"Well, then," the old man retorted, "why is she lying like a stupefied eagle, wingless and insensible?" Having only just caught up with his daughter, he berated himself for leaving. He worried that his absence might have weakened her resolve.

"Where are we?" asked Lala, looking up at the growing crowd of mourners. "What event is this?"

"The funeral of Sara's son, Memín."

"And the other? Where is *he* now?"

"Miguel is there, in this peculiar memory, standing at his mother's side."

Lala looked through the crowd until she spotted him, an eleven-year-old standing close to his mother and looking up at her face as she sobbed savagely. As other relatives moved in to console her, she turned from her son to fall into the arms of her husband. Losing sight of his parents in the crush of people, Miguel edged carefully away, studying the scene from the shade of the elm tree, where his older brothers had gathered in troubled silence.

"This is bad," said his grandfather from his post beside Sarita. "No one is attending to the boys. Yes, they are almost grown, except for Miguel, but this is a heartbreak for them, too. How is it that we neglect the innocent, the uninitiated, in our selfish wish to grieve?"

"Oh, they are initiated," the redhead responded, anxiously rubbing Sarita's wrist. "They have already memorized the script to this piece of human theater. They will survive, of course, by donning their costumes and shouting their well-rehearsed lines to the balcony, just like everyone else. To tell the truth, this is what makes me enthusiastic about humankind. Mindful drama."

Don Leonardo looked at her, astonished. "Mindful?"

"Just look," she said. "You are a great one for looking."

The two of them turned to watch the assembly of mourners. Everyone was now crowded around the grave site in a tight circle—men, women, small children, and bewildered teenagers. Sara, the grieving mother, was at the center. A priest could be heard speaking, but he was barely visible within the throng. Then, after a few moments, even his words were lost, for a wailing rose from the group that was both chilling and disquieting, a sound that obliterated every other sound. Rising from the initial soft moan of one woman's grief, there came a chorus of moans that grew and grew until it felt like a torrent of sorrow, the hymn of a thousand bereaved mothers. Beneath its refrain thrummed the resonant murmur of men, comforting and consoling. The noise swept skyward, up and around in random circles, until it finally reached a crescendo and plummeted to Earth. Up and down it went—swirling, spiraling, plummeting. In the midst of its fury, the priest shouted out, inviting the bereaved to offer parting gifts to the deceased—flowers, notes, rosaries. As the mourners began performing their ritual farewells, the sorrowful background voices began to falter. Wailing settled into whimpering. Finally, the cacophony faded to scratchy silence, like a musical masterpiece lost in the final grooves of an old phonograph record. The funeral was over, and the crowd scattered onto the grassy hillside in separate little bunches, each one advancing toward a waiting car.

Throughout this remarkable scene, little Miguel stood by the elm tree where he had earlier gone to join his brothers. After the brothers had joined the group at the gravesite, Miguel remained by himself, watching and listening. Don Leonardo kept his attention on the boy; he followed the whimsical patterns and images that moved through the youngster's mind. The child was seeing the drama—the great skirmish of emotion that was playing out in front of him—without submitting to the spell of it. As Leonardo dreamed with the boy, he began to relax and to remember, his mouth curling into a sly smile that flitted across his face and found refuge in his knowing eyes.

The death of my older brother was a devastating event for me and for the whole family. He was nineteen years old, and already a husband and a father. Of course, he was still a child to most of the adults around him, and certainly in the eyes of his mother. His death came by surprise, as it does when it touches the very young. Then again, young men seem to woo death like zealous lovers. Memín drove fast, and with reckless pleasure. At nineteen, young men are gods; we are immortal, because we say so. Never mind those who worry over us and who would give their lives for us. And yet, at nineteen Memín was the head of his own family. His young bride was pregnant with their second child. He had already accumulated heavy responsibilities, even as he careened headlong into manhood. Before he could reach it, however, he was killed at the wheel of his speeding car. His little family was with him, and thankfully they survived. In that sense, he lived on through his children, but the brave and blazing light that was Memín went out forever.

By the time I myself was nineteen, I also was too arrogant to listen, and too full of life to respect the nearness of death. In those heedless years, I drank too much, partied too much, and eventually pushed fate against a concrete wall in my merry insolence. I would have courted danger to the point of death, like my oldest brother, had something not prevented me. But something did, and I lived to grow a little wiser. I lived to achieve the promise of wisdom that life makes to every child.

Such wisdom was an integral part of me when I was very young and hadn't yet lost it in the deep hormone drifts of adolescence. At eleven years old, I was still thoughtful. I may even have been wise. I had my dreams, and I had my heroes. Like my other brothers, I saw Memín as an action hero. Certainly, he was always in action; he was always moving, running, speeding, laughing. He chased schemes, goals, girls, and we assumed that nothing could stop him from catching all of them. Wasn't he faster than time? Wasn't he quicker than destiny, and stronger than doubt? Wasn't he the coolest guy we knew? It took a long time after his death to realize that Memín—brother and action figure—would no longer be playing among us.

Strangely, his most lasting gift to me—the youngest brother who played such a small part in his life—was his funeral. My childish thoughts moved toward a kind of wisdom that day. Standing among my relatives, I felt as if I had two families: one was caught up in a scene from one of Mamá's *telenovelas,* in which each character, played by actors with varying talents, wreaked havoc in his own life and the lives of others. My other family communicated through impressions, feelings, and encouragements. This second family might not have existed at all, or they might have been right there, living with me. They might have been my mother, my father, and my brothers, talking to me beneath the noise of their randomly spoken words.

There might have also been a third family with me that day—I could have been sensing a lingering trace of my ancestors. The old ones were gone but not gone, and all of them were wiser than I. Whatever that connection was, I felt I had company that morning when we buried Memín. The mystifying presence of the old ones stayed with me throughout the day, even as we left the cemetery and went home . . . and the family's bitter tears turned inexplicably to laughter.

That's right. As if someone had changed the channel on our tiny black-and-white television, the mood of the group lightened miraculously when the front door opened and women poured into the house to lay out platters of food. Suddenly

I was watching a different kind of spectacle. In this one, the women gossiped, the children played, and after a few beers, the men took turns telling hilarious stories about my dead brother.

I saw how people put on arbitrary faces and took them off—on cue, and following each other's lead. Racked with grief in one instant, they needed just a little encouragement to remove the grief masks and start again with a joke and a smile. They kept up with each other, mirroring responses back and forth, eyebrows twisting and lips moving to the words someone else was speaking. Oh, there was food on the tables, and everyone ate well that afternoon, but I saw for the first time how nobody missed a bite from life's emotional buffet.

And it wasn't all good. With every bite of *biscochito,* they took two doses of poison—feasting on scandal, sharing disapproval, spreading rumors. A kind woman would say unkind things about someone else, inexplicably. A grown man would seem pleasantly congenial one moment and fighting mad the next, for no other reason but that a particular word had been uttered. A word, a phrase, a look, a shrug—what more did they need? I'd been learning how to act this way for years, without realizing that I had become a master at it. It was already easy for me, at eleven years old. It was automatic, but when I watched everyone else that day, I felt the wrenching shock that comes with sudden awareness.

Emotions seemed to be feeding something I couldn't see. They ran unchecked through each human body, causing sickness and frenzy—but for what reason? There was nothing about sadness, anger, or joy that was wrong. I remembered a time in my childhood when emotions ran through me like river sprites—they touched me, changed me, and then vanished without leaving a scar. These people, though, were scarred in ways I couldn't see, and the pain was still being felt. It seemed odd for someone to submit to sorrow simply because the occasion called for it. And a bit later, were they all being jovial simply because it was three o'clock? Would they be terrified by evening, and disappointed by bedtime? There didn't seem to be any rationale for their emotional drama—except that someone, or something, was feeding on the power of it.

In time, an idea came to me. As I listened, and as I watched, I could see that normal emotions turned intense, even vicious, as people were drawn into one story or another. It might be something they were hearing, or saying, or thinking, but the story ruled each of them, and changed them, turning them into hunters, hungry for a certain kind of blood. Sensing, *feeling* humans were being transformed into creatures who devoured human feeling.

I began to play with random emotions, feeling them at my fingertips, as people moved around the little house that day. Without speaking to anyone, I practiced shifting moods and attentions. Sitting on the floor, I steered the subtle flow of emotional energies here and there, getting a sense for how it was done. People laughed, then they cried a little. They comforted each other, and then fell silent. The current would stop, start, then move faster. It would correct itself, making a new pattern, and the moods would shift again. No one noticed the little boy with eyes closed, seeing something that couldn't be seen, as his fingers gently touched the air around him and his expression remained curious but serene.

※ ※ ※

L ook at him. Do you see what he's doing?" asked Sarita, who was sitting on one of the high-backed chairs in the home she'd shared with her husband and children long ago. It was interesting to find her elderly self there, in her usual seat at the head of the table, staring at bowls of salsa and platters of chicken. Sipping a cup of herbal tea, she felt she might recover her strength again.

This kind of scene, where dozens of relatives filled the house and spilled onto the porch and into the street, was as familiar as old shoes. She still loved nothing better than to hold family gatherings at her home—to cook, to eat, and to exchange stories. She could hear José Luis laughing out on the porch, and she felt deeply comforted. These had been wonderful years for the two of them, when the older girls were married and raising their own families, and when the first grandchildren were born. Life in this tiny place had seemed perfect, at least before the accident. After that, it had seemed less safe and less certain.

"I do see what the boy is doing," said don Leonardo, "but I can't see why he's doing it." He went back to picking *galletas* off the dessert tray.

"Of course you can," she said, pointing at the boy, who was still sitting on the living room carpet. "You and I do it all the time. He's watching life flow around the room in ribbons and streams."

"He's not normal; *that* I can say. Maybe before, but not now."

"It was far from a normal day."

Sarita looked around, moved at the sight of so many dear family members. There were nieces and nephews, children and grandchildren—most of them old now, many of them departed. She was one of only a few left of her generation, those who remembered the old times, and yet she had to admit it was hard to recognize many of the people in this room. Had she changed as much as they?

There was an old man sitting on the divan at the far end of the room, a plate balanced on his lap. He was dressed elaborately in a traditional Mexican outfit of flared black pants and a cropped jacket, both studded in silver *conchas*. Beneath the jacket he wore a ruffled blouse, once white perhaps, but now faded to a musty yellow. A large sombrero lay next to him on the couch, grimy with age, its tassels knotted and stained. The old man's skin looked like sun-baked buffalo hide, but his eyes were bright and full of mischief.

"Is that—?" she began, and then stopped herself. "Could that be don Eziquio?"

Don Leonardo gave her a look made of fresh innocence and headed toward the tub of cold beer that awaited him on the porch. Muttering to herself, Sarita rose from the table and moved across the room with slow deliberation, still unsure of her balance. She approached the leathered old man and stood over him as he wolfed down his food and hummed quietly to himself with pleasure.

"Grandfather," she said abruptly. "Why are *you* here?"

The rugged face looked up at her in surprise, beaming a smile of recognition. "Sara! How very old you've become!" he exclaimed, swallowing a mouthful of beans. "I'm honored to be answering the call of my much-bewildered son. He is in need of my advice and expertise, as it happens."

"My father called you? Do you know why?"

"A matter of death and life, I was told," he explained cheerily, ripping the last meaty morsel from a chicken bone. "And he promised there would be women."

"It is a matter of death . . . and life," Sarita said softly. "We find ourselves at the funeral, so long ago, of my sweet boy, Memín. But our purpose here is to save my youngest son, whom you may not remember."

"Of *course* I remember!" he said, patting his lips with a stained napkin. "Miguel Angel! It is for that reason I feel confident there will be women." He peered through the crowd of people. "Which one is he?"

"He is there, on the floor. At this time, he would have just turned eleven."

"Eleven? Is that all? Ah," he said with dismay, hardly looking at the boy. "Then we will have to wait a year or so for willing girls and rhapsodic pleasures. Well, that's no problem; I've got time." He went back to his plate of chicken and beans, looking up briefly when a woman walked by—a gorgeous woman with red hair and eyes as deep and blue as the *cenotes* of his homeland. He looked at her once, then twice, wondering where he had seen her before. No, he had never seen her—and yet somehow they had met. Yes, they had met.

Sarita left him where he was, unsure how his presence would improve the journey. Well, an ancestor was an ancestor, so she wouldn't complain. She'd had enough of this particular memory, in any case. She wanted to be done with it. This sad day, which had been a horrible experience for her then, was somehow made more horrible by its recapitulation. She began working her way to the kitchen, in search of the redheaded woman. They needed to talk. They had a small amount of time available to them, and an even smaller shopping bag.

In her haste, Sarita failed to see Lala milling through the crowd, considering her next move and circling the child who sat on the floor by himself. The redhead had already noticed the old woman, and although she was relieved to see her in good health again, she was tired of Sarita's bothersome questions, so she made herself invisible among the relatives and neighbors who jammed the front room. She liked it here. She liked it when people came together to smoke and talk and spread the virus. Any virus was transformational. Any virus could change the way an organism worked, but this kind of virus changed the human dream. It was a word-borne virus, a virus that inflamed thought and started a fever in the human body. It was knowledge, something that her world would not exist without. She smiled, comforted to know that she lived in that world—a world built out of syllables, sounds, and the strong mortar of belief.

Her world looked the same, felt the same as the physical universe, although some called it a reflection. Her symbol was also a tree, like the Tree of Life—great and lovely and deeply rooted. The roots of life stretched into the infinite and its branches breathed eternal light; but her roots drank from the spring of human storytelling, and her branches bore its fruit. There was no thought, no reality without her, she mused. Without her, there were only beasts in the field.

She could sense the living Miguel in the room, although she could not see him. He wasn't here, where the little boy sat, teaching himself to trace the forces of human feeling. Miguel was near, however, watching and waiting for the right moment to show himself. If he was here, he would be watching this boy, she thought. He would be remembering, and helpfully packing that memory into his mother's shopping bag. He didn't wish to return to the world he'd left, she knew, but he would. He would, because Sarita insisted. He would, because a wise apprentice will honor the teacher, if not the mother.

Lala lay down beside the eleven-year-old that Miguel once had been and looked into his face. Ah . . . that face! And the eyes, hiding a blazing light somewhere in their darkness. These were the eyes of the man he would someday be, the man she had never truly learned to resist.

"Do you know how much I have wanted you?" she whispered to the boy. "Can you see the past and the future of us, my love? Can you see how we will dance together, through a thousand more generations?"

The boy's expression didn't change. His black eyes were focused on things that no other person in the room had noticed. None, that is, but her. Lala sighed, laid her head back on the rug, and closed her eyes. She was recalling the first time she had come to him . . . not just in visions and thoughts, but in the fullness of a woman's body and a woman's intellect. She had waited until he was bored, tired of the same tasteless food. She had waited until he was ready for the kind of knowledge that stirred men into a frenzy. It was only then that she had taken him by the hand and led him back into the ancient dream of the Toltec people.

Like everyone, Lala had been shocked when Miguel left his medical practice and the safety of his books. She worried when he returned to Sarita—who was a sorceress, however she wished to call herself—asking to learn her skills. During those years as an apprentice to Sarita, he had become intuitive, and unafraid of his own power. He was slipping out of her control. Lala wanted him to understand how human beings are connected by words, only words, and to recognize the supreme authority of ideas over human actions. She felt compelled to help him elevate storytelling to its greatest genius, and that was what she did.

Ah . . . Lala knew now where this journey would take them next, and she smiled in satisfaction. She must collect the old woman so they could start up again—so they could witness the moment when Miguel first met the woman who had inspired his storytelling. He had been afraid at their meeting, having recognized her from his sleeping dreams. He wished more than anything to run away from her that day, but he stayed. He stayed, and he fell in love. Yes, that's where they would go next.

She opened her eyes, and when she did, she saw the boy staring directly at her.

"I've never danced with a girl before, but I will soon, I think." He looked around the room and then his eyes drifted back to her. He assessed her, his face flushed with feeling.

"Yes, soon," she whispered. This fledgling student, with his innocent and tender eyes, would someday become the master. It was time for her to shift the dream to her will. This was her chance to move memory's current. Nothing was inevitable, she assured herself, and this dance was far from over.

◉ ◉ ◉

Don Eziquio was on his third plate of food when Miguel Ruiz sat beside him on the divan, his own plate in his hand. Still wearing the hospital gown, he looked more out of place than ever. He was drawn to this time and this place, however. He had noticed his older brothers talking with a few of their cousins in the pebbled driveway, and he was curious to know them again as children; but by sitting here in the crowded living room, he had a view of himself as a child. He smiled at the sight of the boy, sitting there all alone, and re-membered the curious feeling of shock he'd felt when he saw the human drama for the first time. As a boy he had envied adults—not just for their knowledge, but for the spectacular way they generated drama. The adult world had seemed like a soap opera set within a mental ward, and he wanted to dis-cover ways to make it sane again. He had looked for solutions all his life, and at forty-nine years old, he felt he was making progress.

He could see Lala lounging there beside the boy, watching, and casually guiding his thoughts. Would she try to woo him with a story? A revelation?

With any intuitive feeling, there comes the temptation to tell a story . . . to think. While the boy sat there, following the tangible traces of life, she would offer him a story about life. Her stories would seem new, not like the ones he'd heard be-fore, and they would appeal to a little boy's vanity. It would be many more years before Miguel, the man, could appreciate any of her stories for what they were.

Miguel finally took his eyes off the boy and plunged his fork into a dish piled high with food. The two men sat there, side by side, enjoying their home-cooked meal in silence. Neither of them acknowledged the other. Glancing out the window, Miguel saw don Leonardo standing alone in the street, his creamy suit catching the pink light of the evening sky. His grandfather looked like a high-born angel, patiently waiting to see what revelations the moment would bring.

Finishing his third helping of food, don Eziquio finally looked at the man sitting beside him. "Good day to you, sir," he offered grandly. "You are hungry, too, I see."

"Mmm, yes. It's been weeks," answered Miguel through a mouthful of food.

"For me, it has been decades. It seems that nothing ever tasted so good!" Eziquio slapped his thigh enthusiastically with one gnarled hand, causing a cloud of dust to rise into the air. The dust was quickly lured away by the opening of a door, and a wisp of cigar smoke stealthily took its place. He said nothing for a moment, surveying the room, and then he turned to Miguel again and gave him a steady look. "To whom do I have the pleasure of speaking?"

"Your great-grandson, who is not really here," Miguel answered. "Just as you, sir, are not really here."

"Ah!" exclaimed the old man. "Yes, but who, among all the legions of men, was ever really here, my dear *compadre*?"

"You make a valid point," Miguel said, smiling, and they sat in silence again, watching people come and go and listening to the melodic buzz of conversation.

"So, you are celebrating the short life of your brother, I suppose."

Miguel shook his head congenially. "This memory is for my mother, not for me. I'm here to show support."

"That, good man, is not all you are showing," said Eziquio, looking at Miguel's naked legs. "May I inquire, sir, are you in need of clothes?"

"No, I'm fine," replied Miguel, smoothing the gown over his knees and dabbing a spot of blood with his napkin. "I'm in a coma, so it would make no practical sense to get dressed."

"I see," said the old gentleman. "Well, have no fear. Should you eventually die, they will dress you up quite nicely. Look at me," he said, lifting his skinny arms. "I made my exit in theatrical style, would you not agree?" He swept up the sombrero and plunked it on his bony head, sending up another cloud of dust.

"Very striking," said Miguel. He glanced around the room again. This day's memories were about to end, he thought, but the stories would survive to entertain generations. Peering through the crowd, he noticed that the boy was now alone, and he wondered where Lala had gone.

"So many children, all harvested from the rich soil of my loins," the old man commented, nudging Miguel with a bony elbow. "I have done my part for humanity, *verdad*?" he added with a wink. "Who is the little one?"

"That's me," Miguel answered, edging his plate away from the old man's elbow. "This was a significant day for me. Very significant."

"What? Oh, I see . . . significant," the old man said, comprehension shooting across his weathered face. "Significant, yes." He sat quietly for another long moment, frowning slightly as if studying a chess board. There were thousands of memorable moments that comprise a man's life, but only a few that could be called significant. Significant memories were the best foundation for a new and enlightened dream, as both men knew.

He looked at his great-grandson with admiration. "You are playing an intriguing game, my boy."

Miguel said nothing.

The crowd was thinning, and there was a hush in the room. Daylight had yielded to dusk, and the illusory landscape had dimmed. Eziquio, the trickster, lifted a withered hand and rubbed his earlobe. Miguel, the dreamer, laid down his empty plate and gave his great-grandfather a look of unreserved affection. Their eyes met in a moment of understanding. The elderly man started to speak, then pressed his thin lips together. A crooked finger scratched at the white stubble on his chin. He tilted his head slightly, pondering. How he had got here, he could not say. Why *anything* occurred—in human existence or beyond its boisterous perimeters—he did not know. Knowledge had no influence on the dead, in any event. He was free of its rigorous sanctions. He was lawless, a citizen of a land where mutiny had no consequence. He reached toward the man next to him, a man whose eyes reflected the same willful mischief, and placed a hand on his shoulder congenially.

"Rest assured, sir," don Eziquio said, giving Miguel an artful wink. "I am with you now."

6

L ET ME SAY SOMETHING ABOUT THE WORD *SIG-nificant*. It's inevitable that during my time away from the world, Mother Sarita, with or without the help of a guide, would logically pursue memories that most people consider truly important—births, deaths, marriages, and wrenching moments of trauma and triumph. Since the moment of my heart attack she has known her mission. With the help of my brother Jaime, she gathered the family, directed the rituals, and held the intent that carried her beyond her own dream to the dream of a lost son. She remembered the names of her allies, the words that must be spoken, and the prayers that must be said. She let go of the known in order to find me on my way to the unknowable. On the advice of her new companion, she's now in search of the memorable moments of a life she did not live.

Memorable moments are not the same as *significant* ones, however. Yes, it's good to remember the time of a child's birth, another child's death, the first word spoken, an ardent kiss, and a great heartbreak. And yet, woven into that bold pattern of events are many quiet threads of realization—random stitches that change the way a person sees himself and imagines the world. Woven into that pattern, in other words, are moments of transformation. These are moments more significant than college graduations and wedding receptions, which come along at their appropriate time and have their expected effect. Whether such a moment comes as a sudden upheaval or a slight redirection of thought, being blindsided by life changes everything. Unpredictable occurrences shift events in inappropriate ways—inappropriate, because they force us to step off the road that has been charted, even grated and graveled, and set us upon a course with no clear destination. With one altered perspective, rules are abandoned; we are mystery-bound and purposeless. We are spurred into action and unable to give a single reasonable excuse for those actions.

I had many such moments, and a wonderful family unfazed by eccentric behaviors; but the things that change a man cannot be stated, and the subtlest events defy explanation. My mother and father, even in their great wisdom, could not dream the dream of their youngest son, nor could they follow his imperceptible path to truth. They could advise, withhold judgment, and then let go. Through their patience and restraint I learned the value of surrender. Supported by their love, I found the strength to take chances and move my awareness beyond knowledge. With every revelation, my life became more spontaneous and less predictable. Every intent became an act of power. When the heart attack came, I knew I might never again play with that kind of power, and I surrendered yet again.

You could say that I've had a lot of time to consider my return to life while lying unconscious in a hospital bed. From the point of view of those who wait for me, their lives suspended, it must seem like forever. It's been almost nine weeks, in fact, and during most of that time, the prognosis hasn't been good. My loved ones sit with each other in the hospital every day. They worry, they regret, and they often cry. They pray and they plead. They fight against fate, and then surrender to it. Some can't stop laughing.

Yes, there are a few who've taken this ride with me, and so they laugh. Without understanding—who needs to understand?—they feel my excitement and elation. They feel the sensation of freedom I am experiencing while my body sleeps and my brain dreams. They know enough to abandon any hope that I might survive, and still they laugh. Joy has carried them through the long weeks, and we've celebrated together. It's hard to describe my poor body's struggle as it fights for life, with a failing heart and watery lungs; but I've enjoyed the unfolding dream. I am what I've always been, what everyone is and always will be. I am life, aware of my eternal nature and heedless of physical limitations. The view of myself I'd enjoyed for almost fifty years has been obscured, but what remains is another kind of view. This one isn't personal; it's infinite.

It's a view of life's pure potential—well, not so much a view, but a feeling. Without an awake mind to select and to censure, I'm feeling life's possibilities run through me like an ocean current, or the movement of air around a condor's wing. I was always life itself, aware of my entirety while feeling the tug of matter. Free of the dream of humanity, however, I can move my attention anywhere, as life does, with infinite possibility at my command.

Coming back would mean returning to a life of consequences—and yes, there would be consequences. The result of my heart failure and weeks of deterioration would be diminished physical capacity—and ongoing pain. Worse yet, I would return as an innocent who is unprepared for the stark violence of the human dream: I would see darkness in the minds of those I loved, where before I had seen only brilliant possibilities. Having almost forgotten Miguel, I would return to a world where others remembered him too well. Everyone would claim to know him and anticipate his needs—everyone but me. Everyone would have a story about him, and every story would sound different to my ears. Yes, I would feel the consequences of coming back. Humanity would confuse and intimidate me, at least for a while. I would have to learn to walk again, to use words again, and to reclaim awareness. I would yearn for sanity, just as I did in my childhood, and find the same madness. Being finite and frail, I would yearn for the infinite again. I would surely feel the consequences of coming back.

Here, the feeling is good and the view is all-encompassing. My view is infinite—but my mother's is specific, and her determination is resolute. Were I awake, and able to explain my situation with words, things might be different. Applied well, words can tame a restless will and soften a woman's heart. My mother is not so different from other women, any woman . . . and not so different from one particular woman. Lala, the mistress of knowing, has haunted every dream—waking dreams and sleeping dreams—since humanity learned to speak. She has always been available, eager to talk and reluctant to listen. She may be determined to hunt me down in this timeless dream of mine, where words can't serve me and life cradles me in deep silence; but any encounter will prove too challenging. After all, this won't be the first time knowledge has approached me in the guise of a determined and beautiful woman.

I was almost forty years old when I met Dhara. I had already seen her face in my childhood dreams, without knowing who she was. By the time we met, she was a married woman seeking spiritual solace. I was working with my mother in San

Diego. I had left my medical practice and was healing people in the ways of our ancestors, becoming more and more skillful by developing faith in myself. If I hadn't been expecting Dhara, I might have dismissed her as just another student. If I hadn't already seen her in my sleeping dreams, I would have been startled by her disquieting presence.

She stood in the doorway of my mother's small temple, her body silhouetted against a blinding midday sun, her hair blowing lightly in the summer wind. I couldn't see her face, but I sensed her power. We had never met before, but I already knew who she was and why she was here. We had spoken many times in my visions. I was young when I first saw her face in my dreams. Being young, I imagined her to be the angel of death, coming to warn me that my life would soon be over. I didn't understand the true nature of death as I do now, and it scared me. I was sure I would recognize the face of my angel if she ever materialized, and I did.

I perceived the moment of our meeting as one of terrifying significance. Death had come to greet me in person! I knew it was my choice to run, or to face the thing I feared. Sensing that the result would be the same either way, I stayed where I was. There was nothing about her sudden appearance that would have seemed out of the ordinary to someone else, but it was not at all ordinary to me. I doubt that even Dhara knew the reason she came to speak to me that day. Like many women who visited my mother, she was looking for answers. She wanted an enlightened person to reveal the truth to her. She felt a strong need to listen and to learn and a compelling urge to transform her life. More than a prayer or a blessing, she was seeking her greatest challenge. She loved Mother Sarita and felt a kinship with her, but she wanted more. She wanted me. I knew that was true, even if she didn't. Seeing her standing in the sun, I perceived all these things at once. I recognized the angel who had appeared to me in boyhood dreams, and I wanted her to remain a phantom. I felt completely unprepared for the sweeping changes that were coming, changes that I could see rushing toward me in her wake.

None of that really mattered as she stood in the open doorway, speaking my name. Resistance would not serve me now. Afraid or not, unprepared or not, I looked up, I shaded my eyes from the brilliant light, and I greeted Death.

"What is this?" demanded Sarita, looking at the scene before her. "Why are we here, at this place? What about Maria, the wedding, the children?" she demanded, looking accusingly from the redheaded woman to her father. "What about the car accident?"

"Never mind all that," said the other woman silkily. "This is magic in the making." Lala was reassured by the current vision. Dhara, the illusion brought into glorious flesh, had made her appearance; she was standing in the doorway and obstructing the light.

"We must use discipline," fussed Sarita, although her heart was touched to see Dhara again, looking so young and so strong.

"What is your worry, sister?" Lala asked. "Did you not love this woman?"

"Of course I did! I *do*!" the old woman answered indignantly. "But this is not her time. She comes later."

"She comes when he says so, Sarita," don Leonardo intervened. "This is his story, and these are his priorities."

"My priorities," corrected Lala. "This odyssey is mine."

"So you say, my dear, but those are only words." He looked first at her and then at his daughter, who was too weary to listen. "Dhara's presence will change things rapidly, and the lessons that follow are of paramount importance."

"More than near death? More than divorce or family tragedy?" whispered Sarita. Shaking her head, she walked deliberately into the scene where Dhara was speaking to her son. She could feel the sunlight on her flesh, as if she were there with them, and she grew calm. Sarita remembered sending Dhara to him at a time when he was trying to find his way, knowing instinctively that their union would be the catalyst for change. She looked at them now, in the midst of that first, awkward conversation, and was happy she had followed her instincts.

"The other recollections will come," her father said from behind her, "and when they do, you will pluck them like rosebuds and place them in your basket of restoratives with all the rest."

"They are of little importance," the redhead scoffed. "Look at these two now! Remember this, and what comes next. Remember why you love her. Think!"

Sarita moved between her son and Dhara, this American woman who wanted so much to learn, to know, and to conquer her own confusion and fear. Sarita looked closely at Miguel, moved so close to him that she could feel his breath on her. She could see the recognition in his eyes as he spoke to Dhara. It was an act of will that held him to the spot. He wanted to run, but instead he smiled, struggling with his English and refusing to be flustered. This was his moment to rise in love, if he wanted. He had not cared for riches or for prestige. When he had abandoned his future as a doctor and neurosurgeon, he had gone instead in pursuit of truth. He had wanted to find what his ancestors had found, and Sarita was helping him do it. She and his father had shared their wisdom with him, giving him a place to work and to envision; but he soon needed more.

And there she was, standing before him—someone with the power to make things happen in the world-dream. It was true that Dhara would become a daughter to her, and a mother to her grandchildren; but above all else, she was the long-awaited partner for her son. Together, they would discover the wonders of Teotihuacan and all its silent knowledge. Together, they would gather a following of eager students. As a shaman and a master of the dream, Miguel would create huge disturbances for all who followed him. He would fracture the rigid foundations of their reality, urging them to see more, to imagine more. He would change the way they dreamed themselves, and they would meet his challenges—Dhara most of all.

"I loved her because she was an ally," Sarita stated simply. "She was a friend, a conspirator. She was a link between our ancient, secret world and the world of present concerns."

"She was a child of the world-dream, was she not?" said her father.

"As they all were, yes," she said. "She moved events, however. She loved, raged, and cut a path across destiny, like Moses moving the watery reeds."

"Yes!" agreed Lala enthusiastically. "Like Moses, she faced her own dream and transfigured it—but she did it for knowledge."

"She did it for my son."

"She did it for herself, as we all do," don Leonardo said. "But let us proceed. Where does this moment take us, dear lady?"

"To the next moment, naturally," answered the woman with fire in her eyes. "We'll omit the lovemaking this time, if you don't mind, and the tearful tides of yearning and separation." Lala was already growing tired of this moment. They should move swiftly into mythology, the invention of clever minds. "This woman admired your son and earned his respect," she said briskly, "but the heart of this memory does not live with her. We'll move now to the place of the Toltec masters, to the pyramids of Teotihuacan, and to the quaint wisdom of your people!"

Don Leonardo looked at Lala and couldn't help but smile at her performance. So true to her nature, he thought, as he took in the delicious sight of her—so anxious to be seen, to be noticed, and to make her imprint on the lives of others. Let her have her moment. Why not? In the end, love must engulf her. Such was her destiny. Such was Dhara's destiny as well . . . and such is the destiny of all human cunning. The Toltec sages had realized this over two thousand years before: knowledge must ultimately yield to love.

Even this beautiful memory of love's beginning had to yield, it seemed. As Lala spoke her intentions, expecting to conjure the wonders of Teotihuacan, her two companions simply faded from sight, leaving a strange emptiness in their wake, a boundless space that swam with light. All she could discern was a defused glare that reflected nothing and revealed nothing. Lala moved around tentatively, expecting another landscape to emerge. She spun slowly where she stood, her eyes open and her senses alert. She wanted to call out but was unwilling to admit her confusion. After all, she was in command of this dream, was she not?

The light dimmed, then brightened, then dimmed again, leaving traces of images as it shifted. She saw the woman, the beautiful Dhara, still washed in sunlight as she laughed in a doorway. She saw soft shadows behind her, wispy things moving as erratically as restless children. Beyond the shadows rose throngs of people shouting incoherently, and in the uproar she recognized something of herself. She could hear her own voice in the steady roar of human conversation. She could feel the emotional frenzy her own words created. Within seconds, all color and sound drained from the scene before her. The roar dwindled as soon as the human images were gone—and with the loss of noise, all feeling went away. Sarita's wistful sentiments were swept from the air. Dhara's excitement, so real in that moment, had washed back into the past.

And Miguel . . . what of Miguel?

Lala had briefly tasted the emotions he'd felt upon meeting Dhara, and she recalled the event with satisfaction. Once, she had whispered a story to him in childhood dreams, and the story had become this moment of fear in a grown man's life. Such was her power over the human mind! It had been possible to stir him then, and it might still be. She lifted her face, trying to sense him, but her efforts were met with that impenetrable, numbing haze. She was a creature who thrived on human feeling . . . so real, so succulent. Was he feeling something now? Could those feelings be drawn into conversation . . . and could her words catch his attention again? Could she catch it, and keep it?

Lala—the one who called herself La Diosa—held her breath, staggered by the emptiness around her. A moment before, she'd been playing among a man's vibrant memories, and now she lingered in the midst of nothing. She was lost. Yes, removed from the tantalizing emotions of human beings and far from the clamor of human thinking, she was lost. Life lay within the boundaries of this white light, but she would not reach for it. Truth hovered within the sparkling silence, but she dared not breathe it. She longed for reflections, yearned for the old lies— and found she could produce none.

Just as her bewilderment became intolerable, something stirred. She imagined that she heard the faint hum of thoughts again, followed by the inevitable buzz of words. The dream was changing, redirecting itself. Lala exhaled, then gasped, relieved to feel energy return to her. The haze cleared slightly, and through a fine and shadowy mist she saw new colors coming to life.

Without warning, without further spoken command, Sarita had been transported again and was now standing, as Lala had wished, at the top of a great pyramid. It appeared that she, too, was alone, listening to the distant growl of a universe tumbling toward the infinite. As early morning winds whipped at her shawl, she sought out the comfort of a low wall, built in an ancient time and boasting thousands of round, multicolored stones. The surface of the wall was cold. The sun was only now rising behind the eastern ridge, promising warmth, but not ready to deliver on that promise. As dawn spread its dim glow, Sarita realized that she knew this pyramid. It was, in its way, as awe-inspiring as the ones crafted by ancient Egyptians, but there was no African desert below her. Around her, in every direction, lay familiar rolling hills touched by the lush greens of summer. There had been a rain sometime before dawn, and a mist covered much of what she could see below her. She was looking out at the central valley of Mexico, her homeland. She stood atop the greatest pyramid within Teotihuacan, the illustrious civilization of her ancestors, built more than two thousand years ago and known as the city where humans became God.

Yes, she knew where she was, and her mind relaxed. She was cold and uncomfortable, but she was ready for whatever came next. She huddled there, shivering against the wall, until sunlight finally touched her faded pink slippers with its searching, silent fingertips. Grateful for that small rush of heat, the old woman stepped forward tentatively, receiving the first, fierce assault of the sun. She tilted her face toward it, whispering a prayer of gratitude. She wondered at the invigorating power that light had on human flesh and on all the precious organisms of Earth. She turned around slowly, letting the sun's rays seep into all the muscles of her frail and aching body, drinking fully of its warmth. A dream this may be, she thought, but the sensations of the moment were too real to deny. Pleased to be in her own company again, she would wait within this sweet warmth, and she would listen.

It was then that she saw him, standing at the top of the ancient stairs, breathing calmly, evenly, as if he hadn't just climbed to the pinnacle but had arrived there on the dignified wings of intent. His arms were held out from his sides, palms up, and his body was still and calm. This was her son Miguel Angel, the one she was trying so hard to save from impending death; but at this magical moment he was from another time, with a human body enlivened by youth, masculinity, and mental resolve. In this vision, he radiated life. He vibrated with it, causing tendrils of light to glow outward from him, illuminating the world below and coaxing the mist from its hidden sanctuaries.

It had been many years since Sarita experienced this view at the peak of the Pyramid of the Sun. By the time Miguel had begun his many pilgrimages to Teotihuacan, her body was almost too weak for the climb. On countless occasions she had come to teach and to heal, but she could not remember being in this exact spot and in quite this way. She wondered why Miguel was standing before her now, without his apprentices and without Dhara by his side. Slowly, as she began to see things with his eyes, she understood. This was his first encounter with the ruins of this great civilization and the dream of the ancient Toltec masters. This moment was his, and only his. This was Miguel's memory, a gift made available to her by his gracious consent.

The cold had left her. She no longer felt the constraints of an aging body, but was filled with a curious wonder. These were the perceptions of a dying man who might have just begun to fight for his own life. Sarita felt her son's command to her: that she see now what he had seen then, at perhaps the most defining event of his life.

The sun, rising swiftly in front of her, splashed color along the horizon. The mist wore a girlish blush as it danced above the ruins and faded into fanciful skies. Through Miguel's eyes, Sarita saw the splendor of Teotihuacan as she had never seen it before. Below her lay the modest remains of the city's core— its heart and its life force. Beyond the great avenue and all the elaborate structures that once lined it, there had existed countless thriving communities, populated by traders, artisans, laborers, and educators—along with tens of thousands of traveling pilgrims who came from near and far. In its time, Teotihuacan had been the largest city in the world. Now, it was hardly more than scratchings in the dust, an insufficient testament to a civilization that had once influenced most of the western hemisphere. The land that stretched just beyond these ruins had, for the most part, returned to nature, to grass and shrubs and cattle grazing in an open countryside. Yet somehow there was much to see in this rock-and-bone map of an extinct empire.

The more Sarita surrendered to her son's dream, the easier it was for her to see and to imagine. The ruined setting had long been a familiar sight and a second home to her. Now, seeing it as he first saw it, she was so forcibly struck by the masterwork that was Teotihuacan, so surprised at its ambition, that it seemed she had never visited the place at all. She saw it now as an institute of higher learning. *This* was a university campus, so simple and elegant in its structure that it took her breath away.

No, it took his breath away. She felt the thrill that Miguel felt—the first rush of excitement as he made this unexpected discovery in such an unexpected way. He had come here shortly after he and Dhara, both divorced from their spouses, had sealed their union in a spiritual ceremony. They came as tourists, honeymooners, expecting nothing. Growing up in Mexico he had seen many ruins, but never this one. Curious, he climbed the pyramid alone early one morning and watched in astonishment as an ancient dream revealed itself to him. In that moment, he saw not only a piece of human history, but a diagram for human revelation.

This had been a place where ambitious students came to learn, and where a few carefully chosen apprentices were given the chance to transcend ordinary awareness. It may have taken years for young novices to transform. It may have taken a lifetime, but none questioned the process. Living an aware life was their ultimate ambition and their finest art.

Art, yes. After all, *toltec* was the Nahuatl word for "artist"; and the most ambitious spiritual artisans learned their craft here, at this esteemed university. Taking control of one's personal story was considered the greatest human act of power. It must have been a humbling thing to be chosen as an apprentice to the great masters who resided here. Personal impeccability required constant practice, at all times and in all circumstances; and a true master could lift the rest of his community to higher levels of awareness only by becoming impeccable in his own life.

As sunshine flooded the mile-long expanse of temples and stairways, Sarita noticed the emblematic design of a snake built into the city's layout, stretching from one end of the university to the other. The double-headed serpent represented the transformational process that every apprentice was expected to complete. Its two faces personified the gods Quetzalcoatl and Tezcatlipoca, or Smoky Mirror. So many myths converged in her consciousness as she whispered those names! So many different stories had been told, retold, and interpreted by many who considered themselves knowledgeable. Without an alphabet, the people who had created this civilization knew only the symbols that nature provided in the forms of living creatures and in the traits of wind and rain, sun and soil: these were etched into the architecture and stamped boldly unto walls with paint and pebbles.

Wise ones understood the quality of being a snake, a creature fated to push through rock and dust on its legless torso, knowing nothing beyond its lowly circumstances and blind to the infinite world beyond. They recognized that such qualities mirrored the human condition as well. Transcendence required a brave shift in perspective. A metamorphosis of this kind required death and then rebirth. A spiritual warrior would have to enter the mouth of the snake and allow himself to be consumed, surrendering all that he had previously known or understood. If he could endure the challenge, he would emerge as an aware being. He would have achieved the greatest of all masteries . . . the mastery of death.

To her left, Sarita could see the place where this spiritual journey would have started: at the Plaza of Quetzalcoatl, where the serpent's mouth stretched open at the farthest end of the campus. Quetzalcoatl, most often depicted as a feathered serpent, symbolized the world of matter. Imagination served as wings, lifting humanity above its visionless state and beyond the limits of physical reality. The wide avenue that led from the plaza represented the body, and the path each apprentice must eventually take. This road was called the Avenue of the Dead. The god Smoky Mirror symbolized absolute power, the force that created all manifest worlds. Each sacred journey up the avenue ended with him, when the finest warriors became aware of their own divinity.

Sarita marveled at this city where all symbols converged, crashing into each other in explosions of meaning. Venomous snakes, two-headed snakes, feathered serpents—each symbol represented different levels of understanding, and each spoke clearly to the hero who was seeking truth. Humans have always searched for truth, she mused, using the symbols of their time. Brushing aside the smoke and distortion of knowledge, they are able to feel life's shattering message. Like all warriors who came to this great mystery school, her son had lifted himself above the symbols, traveled on pathways of light, and taken flight.

Miguel, the shaman, would take many students to this place of higher learning, creating rituals as he went; but in that first instant, seen from this magnificent perspective, its message was clear. Trainees were challenged to see through the smoke and confusion of symbols. They were asked to experience the freedom that lay beyond death.

As Sarita considered this, the words confounded her. She thought she knew what it meant to be dead. Death was the end of all things, was it not? Death signaled the beginning of eternal glory for every deserving human. To her and to everyone around her, these things had always been certain; but as she felt the morning winds stir, lifting Earth's intense heat heavenward, Sarita saw clearly that death did not mean the same things to her son.

W HO ARE YOU," DEMANDED LALA, LOOKING at the elderly man, "and what am I doing here?"

"My name was once Eziquio," the old shaman answered.

Having little patience for witless humans and their tenacious ghosts, Lala turned away to survey the scene around her. The two of them were sitting together on the top step of a pyramid as the morning sun blazed down upon Teotihuacan, turning the Avenue of the Dead into the spectral vision of a giant snake warming its massive body in a straight line between temples, rocks, and the ruins of an empire. In the distance stood another pyramid, a bigger one.

"You conjured this place yourself," don Eziquio continued, gesturing toward the spectacle below. "We are in Teotihuacan, as you can see—the place where men become gods."

"Nonsense."

"All right," the man conceded, blinking thoughtfully. "The place where a man may learn to see. And, when he finally sees, he sees that he is God."

"He sees what I tell him to see."

"Is that true?"

"Where are the others, old grandfather?" she snapped. "And where is this?"

"The edifice upon which we sit is called the Pyramid of the Moon, and it affords us this breathtaking view of the entire length of the avenue, as well as of that magnificent monument to life, the Pyramid of the Sun." He pointed theatrically. "There."

With that, don Eziquio paused to take in both the present grandeur of the place and its past majesty. However he may have misjudged the timing of this event, he was going to make the most of it. What did he know about time, after all? He was not so much a man as an endless moment, a wave of potential in the guise of something ordinary. As he turned his face toward the ancient city, he felt a spark of inspiration flash within him. Holding himself perfectly still, he could feel the presence of a great dreamer on the brink of revelation. Miguel was near; not here in the flesh, but somewhere in the dirt and air of this place. He cast his gaze to the sunlit pyramid in the distance and smiled. If there was ever a significant occurrence in a man's life, it was occurring to Miguel now.

"Imagine how all this appeared two thousand years ago," he exclaimed, eager to share his enthusiasm, "with temples shimmering in the sun, and giant pyramids gleaming, beckoning pilgrims from all points of the hemisphere—a beacon for wandering souls! Imagine all the brilliantly painted walls, murals, and gilded doorways! Imagine the beauty of this place, the glory that beat from its mighty heart, and how even the memory of it might stir the dead to life!"

Don Eziquio removed his dusty hat and held it against his heart. The glint in his eyes gave evidence to a passion that both amused and annoyed his companion.

"I hardly need to imagine this great city, old one. I was there."

"*Claro*," he responded, casting a quick glance in her direction. "You've held up far better than these temples, *señora*. I sincerely applaud you."

"I am La Diosa," she reminded him, rising from where she sat, as if to impress him with her full stature. The wind caught her red hair and rippled the loose folds of her dress.

"In what dream are you called so, may I ask?" His eyes were clear and innocent.

"Within the deepest dreams of mankind," she snapped. "Now go away. You are not my concern."

"Really? Then what is your concern, my dear?"

"Go!" she shouted. "Leave, harlequin! Find the old woman, if you know where she is, and send her to me!"

"She is dreaming with him, just there," he said, pointing again to the Pyramid of the Sun. "You are not."

"She is what?" Lala blurted. She looked down at the strange little man who was dressed for his own funeral, and then away, at the pyramid that loomed in the distance. It was larger and more magnificent now than it had appeared in the swirling mists of dawn. The sun seemed to have expanded it, restoring to it some inexplicable power. True, it was not what it used to be; but it was, and ever would be, a testament to mystery. "She is . . . there?"

"What similarities does the moon have to the sun, *señora*?" he asked.

"Light. And the power to enlighten," she answered, still gazing at the larger pyramid. "The old woman is there, you say?"

"None," he said simply, ignoring her. "It has no similarities."

Don Eziquio stood up, his slight figure dwarfed by the tall woman beside him.

"You are a fool," she said. "Like those who came before you and those who have followed since."

"Are you sure?" Eziquio challenged. He waited for her to turn to him and then gave her a stern look. The gleam in his eyes might have suggested amusement, but it was impossible for her to tell. When it seemed that the entire world had gone silent again, he spoke.

"You are a reflection, my dear," he said, almost tenderly. "You are the faintest glow, dreaming itself as the sun. You are a false luster. You are a forgery." He paused, attempting to give weight to his words. "I, too, am a trick of light, an impostor. We are neither of us more than crude copies of the one everlasting truth. That is why we stand here, at the farthest end of this avenue, looking at creation from afar, stranded upon this pretentious shrine to matter and to death."

Lala looked at him with contempt and then dismissed him, turning her attention to the Pyramid of the Sun. "Mother Sarita should be here with me," she said. "*She* is what I am. We are the same."

"Not today. Today she is perception itself."

"Words make perception!" Lala argued. She hesitated, as if trying to remember something. "In the beginning, there was—"

"There was no beginning," he interrupted, his expression softening into a smile. "There was never a beginning. You take your meaning from the pages of a book, from the muddled minds of men and women. Eziquio, this wraith you see before you, once thought he had meaning. Oh, he worried a great deal about meaning. The moment he washed those worries from his darkened dream, the sun rose and the world came awake within him. From then on, he has never stopped laughing."

With that, the old man released a loud and happy howl. He started to dance, hopping from one stone step to another like a gleeful child. "Body or no body, I am laughing. I am joy itself, and the sun shines wherever I go."

"The sun is of no use to me," Lala said, ignoring his antics.

"Nor am I," chortled don Eziquio, lifting his face to the sky. "Nor, evidently, is laughter!" He let out another whoop of joy, followed by a belly laugh that made the stones vibrate.

Suddenly the atmosphere seemed to change. Although the sun still shone, a gloom fell upon the place. Thunder rumbled in the distance, but no clouds appeared. Strong winds whipped upward along the steps of the pyramid and tipped the old man back on his heels. He howled again, in surprise this time, and saw a dozen dust devils spring up along the avenue, their funnels rising. Still laughing, he bounded up a huge stone step, then hopped down again, taking three leaps to arrive back at the woman's side.

Lala's hair was streaming behind her. Her eyes glowed red, and her face held a savage look of determination. He had provoked her, and so easily. She was now lost in her own magic. Her arm swept upward, and a cyclone of dust billowed toward the sky, obscuring the ruins below. Daylight dimmed further, and black shadows moved over the morning sun, blotting it from the sky.

Standing together in sudden darkness, the two fell silent. After a long moment, the old man chuckled. "I can feel the foul stench of judgment on my face," he said.

"Good."

"I can sense fear as well, my dear, but it is your fear, not mine." Eziquio laughed again, lightly and appreciatively. "Your power is gossamer, hardly more than a rumor. Can you hear me laughing, daughter of rumors?"

From everywhere, or so it seemed, there came a soft, tittering sound. It was uniquely human, but impossible to comprehend. And then, from the deep recesses of ruined palaces and empty temples, there rose a chorus of giggles. From open fields and rising hills beyond, there came hearty, satisfying swells of laughter. The sounds reverberated from ancient walls and parapets, echoing through the city and pounding at the darkness.

Angered, Lala struck out with an arm, intending to knock the old trickster off the steps and out of existence, but he was no longer by her side. She stopped, listened, and then heard him howling foolishly from the top of their pyramid. She could hear his laughter ringing above her, then below her, and then from all six directions. Nothing could quiet it, it seemed; nothing could end it.

She, who could shift anything—even rebuild this place from its primeval bedrock—could not stop this joyful noise. She could not go to the great pyramid, and Sarita could not be summoned. She heard the laughter rise yet again, and again the night was filled with it. She could hear the old man howling, cackling, and clicking his booted heels to the booms of laughter that now swept the city.

"I will find her where she lives, you miserable goat!" Lala shouted. "I will find her in her own muddy playing fields. I will find her, and together we will return to our purpose!"

With that, Lala was gone from the place. The laughter soon faded, and all was quiet among the ruins at Teotihuacan.

◉ ◉ ◉

What? What? Have they—?" Sarita blurted, her eyelids fluttering as she came awake and stared fearfully around her. "Have they called? Has the hospital called? Do they know?"

· 110 ·

"Sarita? Does *who* know? Who?"

The old woman blinked slowly and forced herself to take in the room. It was her own living room, and her family was still there, still grouped in a circle, their eyes fixed on her. Someone was cradling her head and shoulders, where she had collapsed in a slump.

"Jaime," she whispered, hardly able to hear herself. "Jaime," she repeated. "My son."

"*Si, Madre,*" he answered, his own voice racked with fatigue. "We're all here."

"They must know . . ."

"Know what, Sarita?" he asked, placing his hand on her cheek and drawing her attention back to him. "Tell us what we should know."

"By all means, tell them what they should know, old mother," rang a woman's voice. Sarita's limp body jumped at the sound.

Lala stood with the others in the circle. Her simulated smile was in stark contrast to the worried expressions around her. "Answer them," she said. "They are hungry for words, and all you give them is this nonsense! Drums and songs and visions! Without the words—my words!—there is no ceremony and you have no son. There is no human dream without the words. 'The thirteenth of thirteen' means nothing. 'Life' and 'death' mean nothing. Miguel himself means nothing. I am the beginning and the end of meaning. I am the sum of human knowing!"

Sarita looked at the woman for a full minute, her breath coming in gasps and swallows. Then she turned back to her son. "Jaime, dream again with me," she croaked. "This dream must not run away from us. Chase it! Chase it!"

Miguel's older brother looked up at the group. "Stand together," he commanded. "Stop the drums; empty your minds. Let in the light, and release your love to help Sarita. She needs our help!"

"Help Sarita!" echoed Lala. "Help Sarita!" She snorted contemptuously at the fools who were gathered there, so solemn and so intent. "Does no one think to help *me,* the one who gives meaning to such pointless actions?"

No one could hear her now. A day ago, a week ago, her voice was all they would have heard. Tonight, they rallied their meager energies for the sake of this healer, this woman who was now somewhere beyond thought. Tonight, they belonged to Sarita . . . but after tonight, they would belong to La Diosa again. This she knew.

The sudden eclipse of the sun had instantly turned day into night. Sarita, attempting to make sense of the dreamscape, suspected that this was Lala's doing. Standing atop the massive pyramid, she lifted her gaze to the sky, and through dispersing clouds of dust she could see the universe pulsing above her. Stars blinked and swarmed like migrating fireflies, each one a distinct and brilliant creature. They flowed and swirled by the millions, bunching and bursting apart in gentle rhythms. She had never seen this world in such spirited motion. It was chaos, kindled by the force of purpose. She reached out to touch the grand mosaic of stars and saw Miguel's elegant hand in front of her, tracing the path of the Milky Way. One slender finger pointed to the constellations, each in turn, as if he were just now discovering them.

The two of them were one again. She was back in his dream, standing on the Pyramid of the Sun for the first time, for all time. Past, present, and possible futures swarmed like the stars. A thousand stories met in his vision, but he would tell only one, the story of Teotihuacan and of the wisdom that sprang from these ruins.

"As it is above, so it is below," he whispered to himself, and his words slipped through the cosmos like trivial memories.

"As it is above," Sarita echoed. Suddenly, looking down at the ruins of the city, she saw the Avenue of the Dead light up with stars that twinkled and danced, just as they did in the swirling heavens. The lit-up avenue raced across the darkness with the same bold exuberance as the Milky Way. The two mirrored each other, like parallel highways cutting through a nighttime desert—never meeting, but always together and forever the same. "So it is below," she finished.

She understood in that moment the task her son had set for himself. He would be the messenger and the message. As it was in life, so it was in matter. As it was in all creation, so it was in him. Every object and creature of the universe was a copy of life, the one living being that contained it. She knew no language that could truly explain this, but she had taught her son to move outside the realm of words. Words were meant to serve the mystery, but they were forever deciphering riddles instead.

Words couldn't attempt to explain the map that unfolded before her. Every plaza within her gaze had a name, snatched from Miguel's imagination. Every temple drummed with the noise of ancient rituals, and he claimed each ritual as his own. There was a place for death, burial, and rebirth. There was the large marketplace, or *mitote,* symbolizing the noise and frenzy of the human mind. Further north there was the place of women and the place of men, two temples reflecting each other along the path to realization. At the head of the avenue was the Palace of the Masters. The ancient academy seemed to open its secrets to Miguel; and the dream of a spiritual civilization, silent for so long, began to take shape before his eyes.

As Sarita marveled at what Miguel was seeing and feeling, the vision of Teotihuacan shifted and expanded. It now seemed more like the body of humanity, lying in a deep and dangerous sleep. She could imagine it now—heavy, haunted, and trembling from endless nightmares as the stars sang and Earth spun to their tune. Humanity was starting to stir after countless centuries, and to rouse itself to wakefulness. Miguel was the piper, artful and shameless, returning at last to sing humanity awake with a lively ballad. He came to sing about the wonder that occurs between life and death. He came to sing about awareness, and every rock and stone would join in the chorus!

All that was left of Teotihuacan was ruins and rubble, but it was now coming alive under Miguel's gaze. It was shifting again toward its destiny, offering comfort and inspiration to future pilgrims. Each temple step would erase the past and reveal the eternal present. Stone walls would take human pain and turn it into revelation. In the high meadows, cicadas were singing with the stars, their wings fluttering tiny benedictions into the night.

Sarita was stunned at the spectacle. She felt Miguel's heartbeat drum within her own body, felt his pulse quicken. His eyes were showing her the universe, describing life in all its wondrous subtleties. Then, little by little, the brief night of eclipse lifted. Stars dimmed, and the sun slipped its bonds to flood the city with light again.

With an almost explosive force, sunlight struck her, its rays beaming from one direction, and then another. One catalyzing ray hit her between the shoulders, leaving her breathless as it cast a long shadow in front of her, where she and Miguel dreamed as one. The shadow ran down the pyramid steps to the plaza far below, taking the form of her son as he stood at the structure's pinnacle—his arms out, and with the sun's brilliant flare creating an aura around his head. The same outline already existed in the design of the pyramid's stairs. From the top, looking down, two stairways were joined. Two legs led to the torso, and then to the outstretched arms. The head, and the orbiting aura around the head, were built into the plaza. What kind of dream was this? How old was this symbol, that it could be sculpted onto edifices so ancient, so forgotten, and so far from humanity's point of origin? This proud civilization had existed five hundred years before Christ, and yet these steps, this pyramid, seemed to echo his story.

To Sarita, deep in the memory of this moment, humanity was dreaming itself anew, its nightmare almost over and its providence made clear. Yes, it would awake—not in one human lifetime, but in time. She reflected on the darkest points of human existence, and how one mind, one presence, could change all humanity. Enlightened men and women had come before and would certainly come again to soothe the human dream and challenge each new dreamer to turn his or her life into a work of art. It was time. Again, it was time.

8

I 'M REMEMBERING MY FIRST MOMENTS ON THE PYR-
amid of the Sun, seeing the splendor of an old dream, and
feeling a strong desire to bring it back to life. What if the
vision of the ancient masters could be shared by wisdom-
seekers in these times, so many centuries later? It seemed to me
irrelevant that this wisdom was born out of the Toltec culture
during a forgotten era. This was human wisdom, shared by
messengers around the globe and throughout history.

What Sarita is experiencing with me now is the beginning
of another phase of my life. I left medicine to find out what role
the mind plays in our suffering. A few years later, looking at
the ruins below me, I recognized the human mind as a virtual
maze. The great city of Teotihuacan was designed to reflect
that, with each step and corridor representing the traps, the
detours, and the monumental achievements of our awareness
process. For Toltec apprentices of old, Teotihuacan was a place

where they could go to gain the highest form of education and discover the truth of themselves as God.

In that first moment of insight, I saw that it was a simple thing to escape from this maze. Unlike labyrinths built of walls and shrubs, we can step out of human thinking at any point. We can see ourselves as the symbol-makers, and feel the force of life beyond our words. Identifying ourselves as that force, we can change the direction of the stories we tell. We can let go of the beliefs that move us so far from our own authenticity. We can be supremely generous to our bodies and to the human dream.

I realized, as I gazed for the first time on the wonders of Teotihuacan, that by working with apprentices, I could learn much more, and quickly. As I rid myself of beliefs and expectations, my awareness expanded. If I could do it, surely they could. Staying close to Sarita now, I can share the most personal moments of that experience—the transformational moments, one might say, of this warrior's life. She may understand and be moved, or she may remain unwavering. In any event, this reflection is my lasting gift to her. While I can, I will dream on.

Between one heartbeat and the next, a man can dream his entire life; tomorrow will open up in front of him and yesterday will retreat into time's abyss. There is no better place for personal memories than obscurity. Once lived, the past has very little value. And yet we carry its lifeless body into all future moments, allowing it to crush us with its weight, to identify us, and to speak for us. Even the most capable adults seem reluctant to make a decision without first consulting the past—the corpse—and listening to its endless rebukes. A wise man will ignore such counsel and observe the world from an infinite perspective.

Since early adulthood, it had seemed obvious to me that there was an immense amount of information available in the present, and that a great many agile shifts were needed to move me at life's velocity. When I no longer felt bound to the past, life became increasingly effortless. Gone were the daily doses of guilt, and memory's ceaseless distractions. Life poured into every eager, empty moment, and moments became swift and changeable. That's how it was for me.

During my weeks in a coma, I've felt no pull from the past. The events of my life have become Sarita's affair. I am real to her, and so are those memories, but they and she both live in a landscape of imagination. Detached and attentive to life's command, I feel only freedom and the unreserved love that arises when fear has exhausted itself.

Death means something specific in my imagination. It means matter. Matter is conceived and then it is born. It grows, it multiplies, but it will certainly end. Matter needs an outside force to move it and, once it's in motion, to stop it. That force is life.

From this point of view, it makes sense to identify matter as death—a substance that requires life's energy to create it and to animate it. Life makes it possible for us to move, breathe, love, think, and dream. At the precise point when matter, or the human body, cannot support the force of life—because of an injury, sickness, or deterioration—it begins to decay. The power that gave birth to the body then consumes it. Death surrenders to life, not the other way around.

Physical death is a homecoming. Light is expanding into light. Energy goes on endlessly, shifting intensity and never stopping to rest. As light, it informs all universes, from the bewilderingly small to the unimaginably great. It pounds into matter and radiates outward with unyielding intensity. It fills all objects, as well as the looming mystery between objects.

Within the space of a heartbeat, a person can see all of this, remember all of it, and follow the power of intent into the next dynamic instant. What is intent? It has nothing to do with a mental process; it is not the same as intention. We have the intention to meet a friend, buy a car, or begin a career, but intent is the force of life that we are. To feel that force, we must first realize that we are life. We are the power that guides us, keeps us, and continually saves us. To use that power with awareness is the work of a true seer who becomes his own savior and a comfort to others who share his dream.

Seeing is total awareness in the present moment. To use that awareness to control the story we tell about ourselves is our greatest act of power as humans. Sarita is sharing that realization as, in her trance, she sees me experiencing Teotihuacan for the first time. She came to experience the truth of this in her own life, too. She was an intuitive woman, a healer who could find the place in someone's body that housed poison, caused pain, or needed extraction. She would often fix ailments that medicines had failed to fix, remove problems that surgeons hadn't been able to reach. She had her ways, and they worked. For many years, I borrowed her rituals and brought life to the old symbols, but there came a day when the words finally yielded to intent, and the symbols were exposed for what they were. That day came, but not for many years—not until after I had put shamanism to very good use.

There on the pyramid, where the distant past merged with present consciousness, I began my work as a shaman. I had been preoccupied on my trip with Dhara, feeling changes taking place in me and looking inward for understanding. Seeing the city in that moment, and acknowledging it as a great university designed for spiritual realization, I announced simply that I would begin taking apprentices there. Dhara laughed, looking around us at the vast Mexican valley. "Who on Earth would come to this place, so far from everything?" she teased. "Who wants to be guided by a man who's lost in his own world—so moody and silent?" She had a good point, but a month later, I made my first official trip to the site, and sixteen students joined me. The month after that, almost twice as many went. We began regular pilgrimages, and the teaching grew from there.

When I met Dhara I was already a father, divorced from my wife. I was embarking on a different path, one of discovery and personal challenge. Her meandering path now took a similar direction. Having raised four children of her own, she felt she had a message to give the world and yearned for the wisdom to do it well. We loved each other but were unsure how to build a future together. We satisfied Sarita's concerns by allowing her to marry us in a ceremony of her own design, and then we headed out on our first adventure together, driving through Mexico. It would be a test for us, a way to experience each other outside of our daily routine. It was a honeymoon of sorts, but I also saw it as a chance for Dhara to see, to learn, and to move past her automatic responses.

A relationship is an *event;* two individuals meet and catch each other's attention. Like any event, the length and the quality of it depends on the quality of attention from both sides. A romantic relationship, like any other, can survive indefinitely. Respect is the key. Driven by old habits and emotional dramas, a union will fail, becoming an insatiable monster that eats love and turns it into a thousand investments and fears. What begins with two people and a strong attraction becomes something else: an entity separate from both of them. "Relationship" is an idea that too quickly becomes a demanding tyrant. When that happens, it's natural to ask where those two happy people went—those dreamers who met, kissed, and fell in love. How did love itself become less than the sum of their investments?

In my twenties, I had married Maria, a girl I knew in college. I felt confident that she would make a good mother and good partner. I saw our roles as traditional: I would make a living as a doctor, and she would take care of our home and raise our family. My professional life was full, and I spent very little time at home. Early on, I was busy being a student, involved in university politics and community activism. Later, I was intensely busy as an intern.

For all our affection for each other, she and I excelled at domestic drama. She was often jealous, disappointed, victimized. I was defiant, indignant . . . and victimized. And we were not unlike most couples. For this reason, I have often described marriage as human sacrifice. Vows are exchanged, promises are assumed, and both partners suffer from failed expectations. What can we reasonably promise another person? How do we live, enjoy existence, and somehow meet the high expectations of another? It's difficult to fulfill someone's deepest needs when neither person can identify or understand them. "You are mine!" Maria would say each time we fought, as if repeating that statement could make the words mean something true and irreversible. The red fire that I saw in her eyes on those occasions, I saw in hundreds of other women during my life. Storm winds blew and thunder rolled every time she perceived injustice, every time thoughts drove her to self-pity. What we believe possesses us, and thus too often it overwhelms love—the truth of us.

While Maria and I argued, life provided, as it always does. I was thrilled and uplifted by the birth of our first son, which made all the drama seem worth it. Indeed, we might choose to pay a price for life's miracles, to tell another kind of story, but they are miracles just the same. It's been a long time since I was tempted to suffer for my happiness. I don't pay a penalty for knowledge, and I haven't allowed personal drama to diminish love. I watch humanity suffer in the name of love, and I attempt to deliver a better message. I encourage people to love themselves. I show them how respect can open doors, while fear only closes them. Respect rules heaven, and heaven is within our grasp with every choice we make. No one should have to *earn* respect. We are reflections of life itself. We can respect one another for existing; and we can respect other dreams, no matter how they may differ from our own. Many interpretations of reality exist, and they have a right to exist. We can say yes to someone or we can say no, but we defeat ourselves when we deny someone simple respect.

Respect creates a natural equilibrium, a balance between generosity and gratitude. Receive life, and give thanks by growing and thriving. This process infuses everything with more life. Be generous, and life is generous in response. The balance that comes from giving and appreciating is the proof of love in action.

These things are clear to me now, as indeed they have been for many years. As I imagined my future that long-ago morning in Teotihuacan, it was inspired by my love for the human dream. Even now, as I abandon the dream of Miguel Ruiz, I feel the power of love creating new worlds and new opportunities for awareness. Perhaps Sarita, the master healer, will see this as well . . . and let go.

La Diosa!" Sarita scoffed, now happily reunited with her father. "So she calls herself, although she seems afflicted by the same frights and vanities as all mortals."

"Indeed, she appears much like any other woman," agreed don Leonardo, seeing no sign of their guide. He was looking into an impenetrable fog. He might have been standing in a forest at dawn, where light struggles to penetrate the mighty hemlocks and sound is softly muted. He listened carefully, but no branches creaked and no animals scurried. If he wandered just a few yards, he would risk losing Sarita, so he simply paced where he stood, breaking the mist into swirling, luminescent clouds as he moved three steps in one direction and three steps back.

"Woman? She is no woman!" said Sarita with a click of her tongue. "What naked impudence, to call that a woman!" She knew that Lala was hardly more than a random voice, but she was actually missing her now. Where was she this time? Together they had collected many important memories, but it was unclear where to go from here. She looked around, trying to get her bearings, but all she saw was heavy mist. All she felt was the cold. She was wishing she'd worn a woolen robe.

"Was my grandson really able to silence the voice?" her father asked.

"Yes . . ." she said, pondering, "and no. His mind is quiet, and his peace is enduring, but how can anyone avoid hearing the persistent clamor of opinions around him?" Her nylon bag was getting heavy with memories now, and she shifted it to her other hand. "Shall we go on without her?" Sarita looked at her father, who stared back at her blankly.

"Go where?" he asked. "We have arrived at a fog bank along the imagined pathways of time. Are you still so sure there is virtue in this mission?"

"Virtue? You speak of my son's life!"

"His life is precious, yes," said the old man kindly. "His death may be just as precious—and as revealing. Can you really know which will bring the better outcome?"

Sarita let the bag go, and it landed at her feet with a thump. "Father, if you are not here to help me, I must do this without you."

"Of course I will help you!" he said. "I simply want you to consider your options."

"Is there nothing I can do to inspire your confidence? Will I always be your bewildered little girl?" She frowned into the murky distance, frustrated.

"My . . . what?" he said, surprised.

"How many years have I been a family joke?"

"You, a joke?"

"Perhaps I was less than a good student, but I am a wise woman! Trust me on this!" She was finding it difficult to restrain her emotions. What, she wondered, was possessing her?

"I do. And you are no joking matter," her father assured her.

"How about my cooking?" she snapped with surprising vehemence.

"Cooking? Well, I suppose there are better—"

"My friends! You made fun of them. My husbands! You never stopped laughing."

"*M'ija,* I'm speechless." Lala was nowhere in sight, and yet it seemed to him that his daughter had assumed the very spirit of her.

"Oh, *now* you are speechless! Not so before! You were full of judgments, every day of my life. Have you forgotten the time I dropped the baby?"

Don Leonardo stopped talking, shocked, and then broke into laughter. "Ah, *Dios,* you go back into ancient history. Ha! How unbelievably funny that was!"

"You see? I will never outlive that story!"

"Yet it is you who brings it to mind!" he pointed out, laughing. "You were a baby yourself. A girl of fifteen with a newborn. You turned your back for one minute and she rolled off the bed. This happens to new mothers all the time!"

"Are they ridiculed for the rest of their days?"

"You ran to our house, crying and shouting, 'My baby, my baby! She fell! What do I do?' It was only when your mother and I asked where the child was that you remembered you'd left her on the floor. The poor infant was still there! You ran to our home, even before picking her up to see if she was all right! Can you admit now that it is funny?" he pleaded. "Sad, of course. Strange, yes—but still funny!"

"I was a *child*. I did not—"

"Exactly! You did not know. You were a child having children. If you hadn't been so eager to marry that dullard—"

"There! There it is! I am a stupid, silly woman, nothing more."

"My darling girl," he said, his voice gentle as he tried to soften the message and dispel the lie. "You are the mother of thirteen now. You are grandmother and great-grandmother to many. You are a wise woman and a miracle-worker."

He paused, waiting for her to smile again. She eventually did, although her eyes were burning with emotion. "Let us continue," he said. "Show me how you bring this admirable man back to life."

"The truth is I feel exactly like that idiot girl who ran from her screaming baby." Sarita wiped away a tear and cleared her throat. "Am I being silly, Father? Has my obsession made me unwise? I cannot lose him."

"You will not lose him, child. Let us proceed, keeping in mind that fear makes us deaf to truth." He brushed the dewy residue of morning from his lapel and straightened his tie. "So, where is the woman?"

"The woman." Sarita felt emotion rise in her again as she thought of the redhead. "Why must she manifest as a woman?"

"Why?"

"Woman, serpent, siren, snake. Do women really deserve this shame?"

"Ah, I see." Don Leonardo looked into his daughter's eyes and sighed. "Deserve it, no. In the human dream, there are men and there are women. There is she, and there is he. Who is telling the story of the human dream—he, or she?"

"He, I suppose."

"Yes, he. Knowledge has so often been withheld from women; so they have envied men for having it, no? Does it not make sense that knowledge should appear as a beautiful woman, to be envied?"

"Small sense."

"It is no secret that men have an insatiable lust for knowledge. For that reason, knowledge is often described as a woman—so desirable that she must be wooed and won."

"Father, really."

"Once children are taken from the breast, they are nurtured by knowledge, raised and guided by knowledge. Is it any wonder, then, that knowledge is depicted as a full-breasted woman, a mother, or a protective tigress? It is only poetic! Like any girl, knowledge was destined to lose its innocence. Like any woman, it plays hide-and-seek with the truth." The old man looked at Sarita with bright eyes. "Need I continue?"

"And what of the snake?"

"That metaphor is the most poetic of all! Knowledge cannot take us to true seeing. Women have always been the guardians of that realm: of wisdom. They have been the secret seers, and as every man knows, it is the feathered serpent that can truly see."

"So a woman represents craven knowledge and wisdom—both?"

"I am a man. Women represent everything necessary to my existence."

"Father, you are a man struggling with symbols—and losing!"

"He is a pillar of intelligence," came another voice. "Symbols are humanity's salvation." The redhead came out of the fog, looking refreshed and ready to return to the game.

"So, you have returned, La Pomposa." Sarita sighed. "Just so you know, things were going very well without you."

"La Diosa, if you please." She looked around her. "And things have not gone well. You are nowhere," she said, keeping to herself that this was nothing as bad as her own desolation. "You are lost in fog."

"Symbolically speaking?" quipped Leonardo.

"Symbols serve and enhance humanity," Lala intoned. "Words are like water, like air . . . and like a strong woman."

"Like water and like air," Leonardo said, "they can also become toxic. Like a strong woman, they can abuse their power."

"Please," Sarita intervened, eager to move on. "Quiet your tantrums, both of you."

The two looked at Sarita and then at each other. It was doubtful that either of them could break this impasse, so they waited together in uncomfortable silence. Nothing shifted, no one moved, until the old woman lifted her shoulders slightly and sighed.

Sarita picked up her bag and started walking, her slippers scraping along the rough surface of what seemed to be an asphalt highway. Somehow, by some mysterious intent, they had already arrived at their next destination.

The mist was lifting eerily, and red lights were flashing everywhere. Night was giving way to dawn on an empty stretch of highway, and traffic had been blocked in one direction so that emergency vehicles could more easily reach the scene of an accident. The police were there, their vehicles parked in a cluster as several pairs of headlights revealed a wrecked car. Sarita stopped in her tracks.

"No. Not this." The memory she most dreaded was playing out before her. "Must we go backward? Why are we not following the simple rules of time?"

"Ah!" Leonardo said, clapping. "What rules, my dove? There are many ways to see time. There is 'firecracker time,' as my own father described it, where all events explode outward from this present moment. Then there is time as it is normally perceived—a string of events marching along in predictable order, and according to one's own memory. Then there is—"

"There is only time as people know it," said Lala. "Knowledge proves it is so."

"Time is the desperate device of bewildered minds." Leonardo cast her a rakish look, winking, but she turned back to Sarita.

"Was this memory not on your list?" Lala asked. "You wanted events that defined Miguel's character, and I feel sure this counts as one. Although, looking at it now, I cannot say why." She took in the scene, shaking her head. The car was a shapeless mass of metal, buckled and crushed. "How does almost dying build a man's character?"

"You see, Sarita, there is someone sillier than you!" Leonardo exclaimed with unmasked delight.

"Father, where is he?" she cried, staring at the crash site. "He survived that night, so why is he not—"

"No doubt they've taken him to the hospital, Sarita."

"Well if he is not here, what are we to learn from this?"

"Learn?" Lala scoffed. "Read the signs, for one. Obey speed limits. Follow rules. Respect the written word."

"Rules? Words? You think he considered those topics on such an occasion?" The old gentleman was standing close to her, his breath streaming wisps of fog into the chilly night air.

"You see now that he did not respect the structure," she answered calmly.

"I see now that he forgot to respect his life," Leonardo said.

"I see now that he glimpsed the secret," Sarita added, comprehension creeping into her voice. Still cold, she pulled the shawl around her neck and shivered. "He saw what he was *not*," she said. "His real purpose began here. The old ones brought him to this moment. It may have taken a few more years, but after this he found his way back to me. He heard truth calling. He started to ask, to doubt. He started a romance with mystery, right here. It all started this night."

Lala blinked incredulously. "Romance? Here is where he might have begun an affair with sobriety—and the relationship might have bloomed."

Don Leonardo laughed. "In the dream of humanity, that party of drunken revelers, he is still the sober one." He laughed again, wrapping his arms around Sarita, who was still staring at the wreckage. Knowing that it was not easy for her to witness the scene of another accident, he wanted to give her warmth and encouragement.

Gaining strength from her father's embrace, Sarita stood a little taller. She was finally beginning to understand the importance of this journey, beyond her initial goals and expectations. This was more than saving a dying son. She was finally seeing her son as he was and would soon be. In him she recognized the heir to a lineage that predated the great Eagle Knight warriors and the ancient Toltec people. He was a messenger of encouragement in a dispirited dream. He carried a torch that was sparked before the first fire burned in Mexico, and its flame would ignite all future moments, everywhere.

I remember being at a family dinner shortly before I was in the car crash that almost killed me. It was a typical family gathering, with all my brothers, their wives and children, my parents, my cousins, my uncles and aunts. There was always lots of food at these occasions, and lots of chaos, as children played and screamed and grown-ups laughed at stupid jokes and endless childhood stories. I was having a great time, showing off our new baby to the family and merrily teasing my wife.

Throughout the party, however, I couldn't shake the feeling that this was the last time we'd all be together. My family was precious to me, and the idea that I'd never see them again put me in a strange mood. I felt compelled to give each of them a different kind of attention—talking with each one individually, listening attentively, all the while guarding my son jealously as he lay sleeping in my arms. I was filled with love for my family, and touched with great sadness at the thought that I might never be with them again.

The car accident didn't kill me, but its disturbance significantly altered my personality and my perception. I count it as the most transformative event of my life, for many reasons. At the time I was already a husband and a father. Our second child had been conceived, although neither Maria nor I knew it yet. I was in my last year of medical school in Mexico City, eager to take time off for fun whenever possible; so I went to many parties. I drank a lot in those days and celebrated like a man who didn't know he had responsibilities.

When I heard there would be a big dance in Cuernavaca one Saturday night, I didn't think twice before telling my wife I planned to go. My brother Luis let me borrow his car, and I drove down with a couple of friends from school. The road was a two-lane highway, but wide open, so it was easy to speed. We got there in a hurry. The party was great; we played hard and drank even harder. We drank and we danced, and then we drank and we danced. Toward morning, it was time for everyone to go home, so the three of us piled back in the car, driving through some unfamiliar neighborhoods until we found the open road back to the city.

It was still dark, and I was at the wheel. My friends were talking and laughing, recalling fun moments from the party.

I was laughing with them until I got too sleepy. I must have become very quiet, but they didn't notice. There was no traffic, and I was driving fast. As one of my friends was coming to the end of a long joke, I blacked out and the car swerved toward a concrete wall that ran parallel to the road. I remember nothing more of the evening, or of the crash itself. Since then, however, I've remembered every detail of what I experienced while I was unconscious.

The moment I passed out, everything slowed down. Time became a different kind of entity, serving an unknowable master. I was unconscious, but I was looking at my body as it sat behind the wheel. I heard my friends shout in fear, and although my physical body could do nothing, I felt an urgent desire to help them. With that strong sense of urgency, I also became aware of opening the back door—as if the car were stopped instead of still racing down the road—and carrying one of my friends out of the car and onto the shoulder of the highway. I did the same for the friend who had been in the front seat. After those two were safely out of the car, I embraced my own body, shielding it from an impact that I somehow expected. It came, violently. The car drove into the wall at a high speed and was crushed.

I awoke many hours later in the hospital. A nurse asked me if I knew what had happened, and I couldn't answer. I shook my head. "Oh!" she said, laughing sarcastically. "So you got your friends killed, and you don't even remember!"

I was stunned. I was sickened. In that moment, I wanted to die. And then, seeing my horror, she admitted that she'd only been kidding, that my friends were fine. I wouldn't take her word for it, so she brought them in to see me. Both were uninjured. I was uninjured. My brother's car, I heard, was destroyed.

My two friends were so happy and amazed to be alive that they couldn't stop talking about wonders and miracles. As they talked, I started to feel worse and worse. I would have to face my brother, of course. Even more terrifying, I would have to face his wife. There was no way I could repay them for the car. There was no way I could make up for the worry I had caused my family. Above all, there was no way I could reconcile what had happened to me that night. My friends couldn't explain why they weren't in the car at the time of the crash. No one could explain how someone in the driver's seat could survive such a collision. If I was the one who had passed out at the wheel, then who had protected me? If I was not my body, then who was I?

Nothing in my medical training had prepared me for questions like these. Mine was a scientific mind, and it didn't care for questions that couldn't be answered. That night would continue to haunt me for many years. During that time I became more serious about my work and my family. The things that entertained my friends no longer seemed fun to me. My second son, José, was born that same year, and as he grew up I suspected that the changes in me during and after the crash had somehow affected him, making us more alike. The abrupt awakening I had experienced seemed to have awakened him in the same way—in the womb, where his own journey into awareness was beginning. He developed into an unusual child and became a young man with natural intuitive powers. Many times I have dreamed with him, sometimes thinking I was recalling the events of my life, only to realize they were happening to him. He was in his early twenties when my heart attack came, but still unable to converse comfortably with other people. I always felt that would change, that he would someday speak to the world as a loving messenger, but I also imagined that I would be there to help him.

Now, as I see old memories come to life for my mother, I begin to wonder if she was correct in saying that there is more for me to do . . . if only for my sons. She loves me, of course, and rejects the idea of burying another child, but my children also love me—and count on me to be there for them until they are prepared to dream great dreams on their own. I feel my mother's pain as she sees the smashed remains of the car I drove into a barrier that night. After the accident, she was there to help me make sense of what had happened and urged me to find answers to my lingering questions. She was there for me in my young adulthood, while I explored worlds beyond normal understanding. She was there for me . . . and she is there for me still. Will I be there for my sons? Is this even possible?

It was Sarita's Sunday Dreaming class that changed everything for me after the crash. My wife and I had moved to Tijuana, where I did my internship at the Hospital Seguro Social, and that led to my work as a neurosurgeon, assisting my brother Carlos. I was on my way to having my own practice and a medical career when the Sunday Dreaming sessions with Sarita finally pushed me toward another kind of work. Instead of treating diseases and neuroses, I wanted to know what caused them. This pursuit eventually took me away from medicine and into a realm that was not particularly comfortable for me. Not for me, perhaps, but it was extremely natural to many members of my family.

I had witnessed some marvelous things as a child. My grandfather—and his father, too—was a curious man who did many curious things. I also had an aunt who seemed magical. She loved to have the family over for dinners in the backyard. There was a time during one of those lively gatherings when I was sitting beside her at the table. At six or seven, I was enjoying the funny stories she told about growing up with several mischievous brothers. She nudged me at some point in one story, asking if I would go into the kitchen and get a little blue soup bowl. I jumped up, ran into the kitchen, and met her there, washing dishes at the sink. Breathless and confused, I froze in my tracks, staring at her.

"What are you doing?" she asked.

"I . . . um . . . I'm getting a bowl," I answered.

"Then here," she said casually, giving a blue bowl a swipe with the dish towel and handing it to me. I took the bowl and ran back to the picnic table, offering it to the same aunt, who thanked me with a wink and a kiss! It was a memorable moment, but there were so many such moments during my childhood that they seemed normal.

As a young doctor who had experienced himself as something other than his body that night on the highway, all those little memories came back to me, one by one. I saw them as invitations, reasons to investigate all the things I didn't know about life, and all that I didn't understand about myself. Starting an apprenticeship with Sarita was my best chance at finding answers. So I began all over again, long after receiving my medical degree. I began again, after spending several years as a surgeon. I began by asking the inevitable question: What am I?

9

THIS IS ABSURD," LALA WHISPERED.

"This is *memorable,* goddess," said Sarita. She was slightly more cheerful, now that the vision of the car accident had dissolved. All the same, it was strange to find herself seated in a small room lined with mirrors. She wasn't sure how she'd gotten here, until she recalled the conversation she'd had with Miguel when she first came upon him perched on a limb of the Tree of Life, munching on an apple in his hospital gown. He had spoken then of stars and space, and of how the entire world of matter was a reflection of infinite life. In effect, then, the world was a mirror room. Here she could see it plainly. Her son had seen it—had made the connection with mirrors—during his apprenticeship. Later, he had been able to teach himself in this way, dreaming alone in a mirrored room for hours at a time. He had made a discipline out of this kind of meditation, until he no longer needed the reminder . . .

until he experienced himself, and everything around him, as a reflection of the truth.

As Sarita gazed at the reflecting panels, something intrigued her. "Why are *you* not in the mirrors?" she asked Lala. Although the redhead was now beside her, her body jammed up against her own, Sarita could not see the woman's reflection.

"This is your dream. You tell me."

"This is my son's dream."

"Hmm." Lala frowned. "Then why are we staring at so very many images of Mother Sarita?"

That was a good question, the older woman admitted to herself. Sarita was bewildered by all the reflections of herself—multiplying, proliferating, and creating an apparently endless universe. She could see her physical image from every possible direction. In every mirror she looked old, slumped, tired, and heavy. She usually prided herself on her appearance—she always fixed her hair and dressed nicely—but in these reflections she appeared unkempt. She was looking at millions of disheveled old women wearing worn shawls and sad expressions. She smiled brightly then, hoping for improvement, and a million faces smiled back, lightening the mood within the room. Better, she thought, running a trembling hand through her gray hair and sitting a little straighter.

The space was so small that it must have been a closet of some kind, with eight framed mirrors braced against its walls. There were no clothes in the closet, only mirrors, and someone had placed a small rug on the floor and set candles, now mysteriously lit, on white saucers. Their flickering caused each reflection to come alive. She could feel Lala press against her, one leg overlapping hers and a sharp elbow in her ribs, and began to yearn for a comfortable couch. Illusions should never be this painful.

"You make a good point," Sarita said, her voice echoing off the glass. "This will not be Miguel's dream until he places himself in it."

"There is no room for another person, sister," answered the redhead. "What if your father joins us—and, heaven forbid, the wraith named Eziquio?" She tapped on the mirror in front of her, as if by knocking on the glass she would discover herself there. She tapped, then tapped again, and something in the room changed abruptly. All the reflected images of Sarita were gone, and in their place glowed the countless faces of Miguel Ruiz.

Lala drew her legs back in alarm. Miguel was sitting beside her, his warm body leaning against hers, his presence threatening her very existence. It was as if the dreamer himself was present, not just the memory of him. Whatever he was right now—a desperate body trying to hold on to life or a mind rewriting its own story—the *realness* of him was undeniable. Just by being there, he breathed life into the dream. The mirrors had now become an infinite universe of possibility, rendering knowledge small and inconsequential.

"That's interesting," Lala heard him say. Was he talking to her? She watched Miguel, looking like his younger self, stare into a kaleidoscope of reflections. He smiled, then closed his eyes as if to hold on to the sensation for a few more precious seconds. This is where she had finally lost him; here, in a mirror room like this. The space was decidedly too small. It was difficult for her to breathe now, and it wasn't because they were sitting in the confines of a closet. It was because, in this instance at least, there was nothing to keep knowledge alive. "Extremely interesting," he said, opening his eyes. "Don't you think?"

He was talking to her, but she said nothing.

"You're here," he stated, "and you're not."

She looked in the mirrors and made no comment.

"You exist, and you do not." He paused. "I feel you close. I hear your words echoing in my head like someone else's memory, but I see no signs of you."

"I assure you," she said, sounding oddly timid, "I am real."

"You are . . . and you're not. You imagine you're real, as I once did . . . but you're not."

"Everything moves at my command." Her voice was now belligerent. "The human acts and reacts because I say so."

"You aren't human, however. Look." Miguel was staring at his many reflections, searching for her. "There's only one human in this room."

"I am human intelligence," she said haughtily. "Try to exist without it."

"You're the result of intelligence . . . and you're not. You're human genius . . . and you're not."

"Humans are forever uplifted by knowledge."

"Ah, and never has the voice of knowledge sounded so reasonable!" Miguel smiled, then added, "You're right—and you're not. Knowledge uplifts, just as it demoralizes. It's the cause of all humanity's problems and it's the solution."

Lala wanted to say something in her defense but hesitated, suddenly uncertain. Was he trying to confuse her? She watched his face as he dreamed from the point of view of countless mirrors, and she pondered. Humans were directed by thought, belief, and shared opinion. Beliefs naturally mirrored each other. Thought imitated thought, like the reflections in this woeful little room, creating worlds too numerous to count. Stories multiplied, copying each other with each retelling. How was this bad?

"Truth lives somewhere outside this room, my love," Miguel said. "Here, as in the world of human thinking, we have only echoes and reflections."

"Ah! You sound like the old fool!" she cried, maddened by the memory of his great-grandfather. "I am *not* an echo—and look! I have no reflection, as you said!"

Miguel closed his eyes, as if to imagine her as she was. He sighed. "The only real thing in this room is this human. Knowledge, my love, is a voice, an ongoing story. It alters human reality for the better, yes . . . but also for the worse."

"Your story will be all that survives you, and I am the protector of stories."

"You are the storyteller, that's true. And the truth-seeker. You are the liar, the confounder . . . and you are none of these."

She leaned close to him, whispering in his ear. "I am you. Yes, I am you! Admit it. I am you, and I am real."

"I admit you want to be real. I confess you are not, however. As this body is my witness, you are not." In the mirror, she saw countless images of Miguel placing a beautiful hand over his heart. She felt the hot fires of rage.

"You want truth? You search for the thing that cannot be found, and all the while, I am here, eager and accessible. All the while, I am whispering, beckoning!" Struggling to resettle herself in the cramped space, she found herself crouched directly in front of Miguel. So be it, she thought. Let him look at the face of himself. "With me, there is certainty, there is clarity," she said with conviction. "With me there will always be rewards, without penalty!"

"The penalties are huge," he countered. "Humans pay an excessively high price for believing."

"What is human life without belief?" she asked.

"You have been whispering that in my ear forever," he said wearily. He looked at the place where her reflected image should have been. "There is only life, my love," he whispered, "and infinite points of view. See?" He nodded toward the reflections within reflections around them.

"See *me*," Lala urged, suddenly anxious.

"I do," he assured her.

"Acknowledge me."

"I have."

"Say you are mine!"

"You are mine," he said. "Now go."

Before she could protest, the dream was over, the mirrors were gone, and she had been expelled from the mind of Miguel Ruiz.

Looking back, it's clear to see that working with mirrors was a key aspect of my learning, helping me to understand the nature of light and its reflections. Matter is a creation of the truth, as are all things related to matter. Matter reflects light, so the brain is itself a mirror, sending light's information throughout the body. From that perspective, we can see that the mind, so vital and complex, is but one of its many reflections.

After my apprenticeship with Sarita, it occurred to me to build a space for myself in one of my hall closets, to fill it with mirrors, and there to dream. I dreamed about matter and the human mind. I dreamed without thinking, without my former reliance on knowledge. I dreamed there every day for many months, and the memory of my revelations during that time is precious to me. It's a significant memory; together with other significant events, it has the potential to frame a new dream.

Knowledge, as I began to see, is the main character of anyone's story; it is a reflection of the truth. Putting knowledge in proper perspective requires a change of mind. The mind must be skeptical about itself. It must be willing to disbelieve. It will go on doing its job as a storyteller, of course—but with complete awareness. Without awareness, we obey the rules of knowledge and live according to its terms. Common knowledge, collective knowledge, assumed knowledge—these determine the way we move through the human dream, and they make our most important choices for us. Our best beliefs, like all stories, can be told in different ways and from various points of view. We can also put them away at the end of each day. We can laugh at our own storytelling. We can turn attention away from any belief. If we confuse belief with absolute truth, how will we move to the next revelation? How can we be artists, telling new stories that transform reality?

It took me several years after the accident to put my own knowledge in perspective and willingly open my mind. Nothing in the world of science could explain my experience, so I came back to my family and to shamanism. I became more removed from the certainties I had depended on as a young doctor. I had my parents, my grandparents, and an age-old tradition to guide me in ways science could not. I also had my brother Jaime—the one who had goaded me during childhood, but always took delight in my passion for inventing new games, new strategies, and new ways of looking at life.

I'm pleased that my mother sought Jaime's support during her mission. My brother and I were close in childhood, but the demands of school, girls, different friends, and separate interests pulled us apart. Once we were married, family came to mean something other than parents, brothers, and sisters. Family now meant wives, babies, and in-laws. In time, marriages failed and children grew up. My life eventually resembled nothing that was familiar to me, and Jaime and I were almost unrecognizable to each other. It took years for that to finally change. The change began while we were both learning from Sarita, helping with her healing sessions and dreaming along with her students every Sunday.

The Sunday Dreaming sessions were very much as they

sound. Each Sunday morning for a year, twenty-one students would gather in my parents' home to dream. Dreaming meant many things. It meant sitting still for long hours. It meant asking the mind to become quiet so that the brain could perceive on its own. It meant a different kind of learning, often while the brain was in a trance state, and the result was a growing awareness of reality as a dream.

Unlike our waking reality, there is no structure to a sleeping dream. A trance is much the same; it is unstructured and obeys no laws. On those Sundays so long ago, dreaming usually included enough time to journey into a trance. The trip might be short, or it could last the entire day. My mother was sensitive to each student's needs and respected the individual process. On one such day, with her careful attendance, I slipped into a dream that seemed to last an eternity . . .

As soon as my mind surrendered, I was no longer in my mother's living room. I was standing in a long hallway that ended in an alcove, where torches emitted a dancing light. The walls were of red granite, and upon every wall were carved detailed images of people, animals, and figures too obscure for me to recognize. Countless symbols were set in precise lines and columns, appearing like sacred texts. The depictions were not in the style of my Aztec ancestors. They seemed to belong to a civilization that had existed somewhere else, and to a dream that had given birth to a thousand other dreams. My impression was that I was in a secret passageway, concealed underneath one of the pyramids of ancient Egypt. The minerals in the rocks sent electricity running along my flesh, and exotic spices tingled my senses, bringing impressions of a forgotten human dream.

Even in my trance state, I was very excited. It didn't matter how I'd arrived at this place; it promised great opportunities. My curiosity was intensified by the appearance of a tall man dressed in white robes. He was bald, and his smooth head shone in the torchlight, giving him the look of a divine being. He regarded me for a moment, and then he spoke in a voice that resonated with infinite patience.

"Do you know who I am?"

The answer came easily. "You are a hierophant," I said.

"And do you know where you are?"

This was not so easy to answer. "I am . . . in some sort of school, I think."

"Yes," he said, smiling. "A place of learning. The symbols that you see, the messages that are so carefully carved into stone, were meant to be read by those who see. Are you prepared to see, and to learn?"

He had my full attention now. This dream had no reference in my reality, and yet the situation had become very, very real to me. "Yes, I wish to learn."

"Then you will stay," his deep voice commanded. "You will give all your attention to these engraved walls, and you will not leave this place until you understand what is written here."

Many questions came to mind as he spoke those words, but before I could ask them, he was gone. I looked around, not knowing how to begin, and then I moved through the room, tracing the symbols with my fingers as I inspected each wall closely. My excitement was great, but so was my fear. What if I couldn't read this script? What would happen if I found there was no way to break the code? If my intellect failed me, how could I ever escape this breathless space?

The hall was narrow but very long, and its walls stood twelve feet tall. Every inch of this space was covered in hieroglyphs. Some images seemed familiar and some evoked interesting ideas, but I could make no practical sense of any of it. I was eager to devote my intellectual energies to the job of deciphering the language and understanding its messages, but it seemed the more I tried, the harder the job became. Sometimes patterns emerged, but my interpretations were clumsy and nonsensical. When I tried to apply familiar logic to unfamiliar sequences, the result was gibberish. When I put meaning to each picture, I was left with unrelated ideas. I felt as if I were in Sarita's kitchen as a child, listening to a dozen conversations at once, and understanding none of them. To a small child, adult conversation is just noise, but it can be soothing in its way. It tells a child that people are close by, that comfort is within reach, and that, however noisy it gets, he or she is safe. My ancient hallway was becoming a noisy place, but it felt far from safe. Its walls screamed at me. Its symbols shouted, argued, and contradicted each other. Their meanings overlapped and collided. Knowledge crowed with self-importance, and the crowing never stopped.

No, there was absolutely no comfort in this noise, and I saw clearly how this defined hell in the human dream. I became dimly aware that I was dreaming then, that I was in a trance and my body was resting in a real room somewhere. I was aware that people must be waiting for me to wake up, but awareness did nothing to change the dream. I could have been dreaming for minutes or hours, but I had the strong impression that centuries, even eons, had passed. Every moment seemed slow and endless—time passed, and I had made no progress. The job now seemed overwhelming. The challenge was too great. I started to panic, but with more fear came more confusion. When I finally realized that I had no chance of returning to my waking state, the fear lessened. When I knew there was no way I could capture the wisdom imprinted on these walls, and that I would have to remain there forever, the noise subsided.

Surrendering, I collapsed. Suddenly, all was silence. All was calm. There was no sense of urgency in me, and no sense of time. I might have been time itself, reclining easily while countless cosmic events stirred within me and around me like restless molecules. Whatever the disturbance, my eternal nature could not be changed.

When I dared to look at those towering walls again, it was with different eyes. Now their messages reached into my core. It was as if their essence and mine were suddenly the same. How could I have been blind to this? What I observed was old knowledge, timeless and familiar knowledge, as familiar as humanity itself. This was the history of human knowledge, dating back to the earliest Egyptian texts, sometimes described as the Book of Thoth. The wisdom documented by Egyptian seers encapsulated all extant knowledge and became the foundation for all future knowledge. This is what I saw.

Liberating myself from a desperate need to understand, I could finally reach total understanding. I saw symbols only, and recognized their particular intent. I saw past them, to the minds of the very humans who had created symbols as a way of sharing the truth. In every tradition, in every culture, this wisdom has existed, kept in secrecy by the dedicated few. Symbols change, morph, and expand into spoken theories. Like the sacred Tarot decks, whose wisdom has been diminished by superstition and endless parlor games, sacred knowledge has always yielded to common knowledge. Words and symbols are distortions of something real. I saw all of this, and more. In my wonder, in my delight, I wanted the tall man to return. I wanted to tell him what I had seen, and I wanted him to show his approval.

At that moment, he was there, standing in the alcove, his white robes shimmering like gold in the torchlight. He looked at me for a long moment, his expression unreadable. Before I could gather my thoughts, he spoke.

"You may go now."

Go? That was it? I felt painfully disappointed. I wanted to shout, "But I know the truth!" I wanted so badly to explain everything, but I realized that it was my own approval I was seeking. There was nothing to be gained by explaining to myself what I already knew. It occurred to me then that once words are spoken, the truth is lost. Symbols might be carved into granite and recorded for the ages, but truth will always live beyond the words, beyond the reach of those wonderful artists who drew pictures in stone . . . and beyond the minds of all those who interpret them.

I was awake in the next instant. I could smell supper cooking, and I could hear the sounds of happy chatter coming from my mother's kitchen. My world was exactly as I had left it, it seemed, but I would never be the same.

N OW WAIT," DHARA INTERRUPTED. "WHAT IS IT
we're trying to accomplish?"

She and Miguel were walking toward the great
ruins of Teotihuacan. The year was 1992, and this was the first
journey to include apprentices, some eager and intrepid, and
some haunted by fear of the unknown. As they prepared for
the day ahead, Dhara talked and wondered, asking questions,
but hardly stopping to listen for an answer.

"I love the place, the history," she continued, "and the pyramids! Oh, God, isn't it beautiful?" She opened her arms enthusiastically, taking in the extraordinary view as Miguel listened
wordlessly. "This is a place of power; there's no question. I feel
it when I'm standing here with you. So we'll create the rituals,
and . . . what?"

"What, what?" he said, as she spun away from him.

"What should they hope to get from all this?" she asked.

"Not 'hope.' Far better to be hopeless."

She tried again. "What will students take away from these journeys, Miguel?"

"Awareness."

He took her hand and led her to the steps at the Plaza of Hell, a physical reminder of the great *mitote*. In the human carnival, everyone is selling and everyone is buying. Everyone talks and nobody listens. How could this be the beginning of anything, she thought, much less a journey toward heaven? Miguel looked out over the plaza and saw the small pyramid behind it, dedicated to the god Quetzalcoatl. That was where it would begin, he said. Students would put their private hells behind them and leap toward the unknowable. They would put their beliefs aside and become aware.

"Aware, yes," Dhara repeated. "But . . . how?"

"They will stop believing themselves."

"Well, maybe," Dhara said with growing concern, "but we need to give them something. They need to believe something."

"Are you sure?" he queried. He was still staring beyond the plaza, seeing what couldn't be seen and imagining what the day would bring, given sixteen students and some exceptionally strong intent. It was early in the morning, but students would be gathering here within the hour.

"Without belief . . ." she began, but her words faded into frustrated silence. "Miguel, for heaven's sake!" she insisted. "What will they *do* here?"

"They will see their lies, and they will end them."

"End lies? Like how? Like . . . do we tell them there's no Santa Claus?"

"We tell them there's no *them*. They are a fabrication."

"Today? Right away?"

Miguel laughed and gave her a reassuring hug. "Today,"

he said, "we'll start with hell. We'll begin with the *mitote,* the dream of humanity. They'll have the chance to admit they're in hell and decide if they want to leave."

"Okay . . . and how do they leave?"

"By understanding what keeps them there."

From where they stood, he could imagine the journey so many apprentices had taken through the ages, guided by impeccable masters. This was where young initiates had faced their demons—their fears and their worst judgments.

"What keeps us in hell?" Dhara asked.

"There are many justifications, but no excuses," he answered. "Humanity is addicted to suffering, and there are a thousand ways to feed that addiction."

Dhara's frown showed her skepticism. "Addicted?" she asked. "So you're asking them to end their addiction, to face their excuses. You're inviting them to stop lying to themselves."

"Exactly."

She took a conscious breath, uncertain what to do next. "Shall I start?"

"Sure! Go ahead!" Miguel said enthusiastically. He left her where she was and took a seat at the far end of the steps.

Dhara moved away, choosing a spot that afforded her a bit of privacy, and meditated on her life. After a few minutes, she looked over at Miguel, her expression clouding.

"Is this like a confession?" she asked.

"Kind of," he said. "You're confessing to yourself." He considered her for a moment, then made a suggestion. "Why don't you pick up an object—a rock, something like that—to serve as an icon. Let it symbolize your connection to hell . . . your hatred, maybe, or your fear, or your guilt. Maybe your pride," he added with a grin.

Surveying the ground at her feet, she found a plastic bottle cap. She held it reverently for a moment. "Okay," she said at last. "I judge. I can't help it. And I feel judged all the time. Do I have to tell you all this?"

"Tell yourself, Dhara. What makes you want to judge? See how judging hurts you and those people you love."

Dhara turned from him and closed her eyes, holding her bottle cap as if it were a sacred object into which she poured all the poison that had been generated by a lifetime of fear. After a few minutes she started crying softly, but Miguel stayed where he was. Something that had begun as a game, spurred by her curiosity, would become an act of transformation. He would not interfere. The moment went on, with more tears, more sighs.

When Dhara had composed herself, she found her voice. "And now?" she asked steadily.

"Now forgive yourself," he said quietly, "because you didn't know what you were doing to your human."

"I didn't, until now."

"Now you have awareness," he stated. "Now you feel responsible to this human. Going against yourself is the only sin. Even by hurting someone else, you hurt yourself." He looked at her, and then went on. "This process can be called repentance. The sacraments were intended as an important tool for human awareness—but they've lost so much power and meaning. Make them powerful for you."

Several of the apprentices were gathering on the steps above. Observing the scene in front of them, they silently took their seats and waited. Dhara was crying again; this time her sobs were more audible. Miguel stood up, walked toward her, and waited. In time, she grew calm.

"Now," said Miguel softly, "decide on the kind of penance you want to pay. It could be compassion for someone, instead of blame. Maybe you'll decide to be generous instead of critical.

Maybe you will agree not to reject yourself. When it's done, bury the icon. End it."

"This is so painful," she whimpered, as if in agony. "I feel like it's killing me."

"It's killing the lie of you," he said, watching her stand up and descend the last few steps. She found a stick and scratched a hole in the dry ground with it. Then she placed her bottle cap in the ground and covered it with dirt, muttering a prayer of gratitude. What had she buried? A little lie, repeated again and again—or had she buried Dhara? Relief had come, either way. She looked up at Miguel expectantly.

"Go ahead. Find the next thing," he said, pointing toward the open plaza. "You're not out of hell yet."

That wasn't what she'd expected. "Not again," she pleaded. "No more."

Miguel indicated the growing group of students behind them. "They will have to do it many times before they can even *see* the gates of hell," he said, so that everyone could hear. "You will do it many more times, too. It's time to break the human form—to challenge your entire belief system."

"Let me help them now."

"I'll help them," Miguel said. "You keep going. You have a good life, but you carry hell with you night and day. Like everyone else, you make yourself pay a thousand times for something you did once, and long ago. You make others pay . . . for your fear, for your knowledge." He hesitated, then gave her a hard look. "Will our love have to pay, in the end?"

She looked at him with a flash of anger and then relaxed. "I want to see what you see," she said.

"So keep going, sweetheart."

He watched her walk away into the plaza, and into the hell she had labored a lifetime to create in her mind. It was a first step. There would be many others, he knew. There was so much she would discover once she decided to live in heaven, where respect ruled and happiness came effortlessly to the human being. She would discover that leaving hell also meant leaving her favorite stories. Even heaven appeared less desirable once people saw that their beliefs were at risk. They were willing to doubt their true nature, but not their knowledge.

It was right that Dhara should be the one to lead the journey. She was the one who had sought him out, who had emerged from his sleeping dreams into his waking reality. She was a woman he could walk with. She was his woman, and her life-long happiness was at stake.

Miguel instructed his students to do what Dhara had done; then he strolled the length of the plaza to view the Pyramid of Quetzalcoatl, a much smaller version of the two pyramids farther up the avenue. He stood on a high platform that faced this pyramid. Between the two structures was a deep gap. He could picture it as it had once been—a pool of deep water that had filled the space between the plaza and the holy temple. On moonless nights, apprentices would jump into the dark chasm, expecting death.

Dying was a necessary part of being reborn, he knew. Along with many apprentices, Dhara would die today. She would die a little more tomorrow. With each surrender, she would die to the existence she had known. Miguel knew she would make a good warrior. What he didn't know was how fast (or slowly) the process would run, or how much personal attention it would take. In time, she would learn what the ancient masters had learned. The minute we take molecular form, we cease to know ourselves as infinite. With our last breath, we die to the finite and the temporal. In the dream between birth and death, there is only awareness.

Awareness came to the apprentices of ancient Teotihuacan one death at a time, beginning with a leap into the dark. By accepting the possibility of physical death, they showed their commitment to a process that would last the rest of their lives. Their apprenticeships took time—years and decades. They jumped. They jumped again. Even without the black pool of water, they faced fear constantly. Fear gradually became a more subtle demon. What all people feared most was the mystery within. Far better to risk physical death, most have thought. Far better to jump off a ledge and into black space.

Great spiritual warriors look deep into the psychic chasm and, confronting their worst fears, find renewed awareness and ultimate peace. Miguel looked again at the pyramid, and imagined the temple that had once sat atop the grand structure—with a hundred stone serpents and sea creatures defending it from the forces of evil. Here, the journey would feel safe, he thought. Now, he would begin the tradition again. Apprentices would find him. Seekers would come to this place in greater numbers, eager for heaven and prepared to not know themselves when they left. For the apprentices of ancient times, the process of transformation would have lasted years, perhaps a lifetime.

He expected the journey to take four days.

Dreaming of Dhara, I am reminded of the idea of spiritual warriors. A warrior is one who engages in battle. He or she is a soldier, a fighter. Every warrior wages a war against something. As humans, we are familiar with war as an act of violence and aggression against other humans. Faction against faction, nation against nation, war is meant to resolve conflict; but, of course, it is the ultimate conflict. Killing incites more killing. Defeat provokes retaliation and vengeful uprising. War, as we know it, resolves little in the long run, but there is another kind of war that changes human behavior. If every man and woman had an appetite for this kind of war, where not a drop of blood is spilled and no humans die, aggression between nations would make no sense to us. It's a different kind of mind that looks inward and perceives the need to fight a battle against its own long-held beliefs and judgments. It's a rare mind that looks, sees, and then decides to wage this war—the *last* war.

A spiritual warrior is something unique. All battles rely on weaponry; and anyone can see how we defeat ourselves—and then others—with the weapons we call words. Words don't need to be spoken to devastate. They have only to be thought. Words band together to form opinions, which strut along the back roads of our minds, inciting doubt and controversy. Opinions rally toward a cause—a belief. Beliefs that are shared by a community assume the status of truth. Without awareness, the war of ideas becomes a war against humans, but the war of ideas can be fought and won within each of us.

Ideas and innovative thinking are wonders of humanity; but when ideas battle against each other, perfect minds are put at risk. Imagine if every human had a disease that showed itself—a severe rash, for instance. If everyone walking on Earth had oozing, bleeding sores on their bodies, on every inch of skin, we could reasonably say that humanity was sick. The wounds that corrupt human thinking cannot be seen, but they are so real and so prolific that we can reasonably say humanity is sick. Childhood traumas and disappointments continue to cripple otherwise healthy adults. Guilt, shame, and tireless blaming turn people into victims. Constant judging makes people mean, and thoughts of injustice make them perpetually angry.

All political and cultural wars begin in one person's mind—and that mind wants company. We want like-minded friends.

Waves of aggression are most likely to arise from peaceful home fires, where a thought begins, then spreads to a conversation, and then becomes a rallying cry. One person's psychological wounds spread to another, and then to many. The wounded mind, however, isn't quite as wounded as it likes to think.

There is nothing either good or bad, but thinking makes it so. I don't know if Shakespeare believed what he was writing, but he could have gone further. Nothing is actually good, bad, right, or wrong. We were told by parents and teachers to respond to life according to their ideas of right and wrong, and we are still bound by those rules. Without those boundaries, we are free to experience and to see. Without the intrusion of opinions, it's possible to respond authentically. Whether something is good or bad, whether someone is right or wrong, is beside the point. This is. They are.

Everything is. Life is. You are. I am. We can choose to participate in someone's dream or to turn away, but there is no compelling reason to judge. *What am I?* I once asked myself. I don't know and I will never know. I have no opinions about myself, although clearly everyone around me does. People have opinions and judgments that have nothing to do with Miguel or his dream. Still, I exist in their imaginations. People assume they understand the motives that drive other minds, and the feelings that exist within other hearts. Their dreams are populated by imaginary friends and enemies, but there is only one character that can be understood and transformed—the main character of their own story. Every person is capable of seeing himself, hearing himself, and modifying what he sees— not because he sees imperfection, but because he sees infinite potential.

My early pilgrimages to Teotihuacan sharpened my awareness of the many consequences of domestication, the early training we received from our parents and our community.

Each of my followers had survived childhood and become a functioning adult. Every one of them was intelligent, with a healthy brain and body. They had convinced themselves that they were sick, however, or had been wounded by childhood experiences. It seemed obvious to me that another kind of domestication was necessary. They didn't need another story, and yet stories begin the process of change; so I gave them a better story. I gave them wonderful mythologies, some Toltec in origin, but not all. I explained the stages of human awareness as they relate to the shifting intensities of the sun. The First Sun, according to the Toltec masters, entered with the dawn of humanity. Over millennia, as the sun changed, the dream of humanity experienced five distinct changes and evolutions. Having just entered the era of the Sixth Sun, we now look forward to expanded understanding and awareness. As each person undertakes to change his or her world and personal level of awareness, the collective human dream will shift.

I told my apprentices about the Toltec people, who had endeavored to be artists of life. I told them stories that encouraged them to love each other and respect each other's dreams. Everyone looks for someone to love them, even when they hate themselves. They wait for lightning to strike, for the phone to ring, for a knight to arrive in shining armor. When people are disappointed in love, they think they're unlovable. They feel deprived and always hungry. This hunger makes them desperate for any offer of love and attention, and susceptible to all forms of mistreatment.

To encourage my apprentices to love themselves, I told them about the "magic kitchen." If each of us lived in a house with a magic kitchen—where the cupboard was always full, the refrigerator was well stocked, and supplies were endless—then of course we wouldn't go hungry. We wouldn't even be afraid of going hungry, wouldn't need to pray for someone to deliver food. So it is with love. If each of us could love ourselves as we wish to be loved, then we wouldn't let our hunger for love make decisions for us. When we're not desperate, we can refuse seductions—even those that arrive in shining armor and bring along a delicious pizza!

Over the years, I changed the stories, replaced favorite mythologies. Old beliefs would be challenged. New beliefs would be discarded. Having attached emotionally to a certain story, my apprentices would have to let it go and move on, abandoning a belief in the process. Attach. Detach. Attach again. Detach. Believe. Let go of belief. Over and over, the process was repeated. Students were asked to tell the story of their lives again, then again and again, until the stories had no emotional charge; until they sounded like all other stories—familiar, but not compelling.

I gave my apprentices frequent opportunities to doubt. "Don't believe me," I would always remind them. "Don't believe yourselves," I would add, "and don't believe anyone else—but listen!" No matter how the myths changed and my stories evolved, that lesson remained. It is at the heart of all wisdom. It's easy not to believe other people when you no longer believe yourself. Every mind creates reality based on its unique reservoir of knowledge. The resulting reality is that mind's personal dream. By listening, we can hear that little dream speaking. By listening, we can recognize the beliefs that give it life. There is only one dream we can change—and it is our own.

More than lessons and mythologies, I introduced my students to a different kind of storyteller. I was someone who had taken charge of his own story. I was someone who was no longer susceptible to knowledge, even my own. I was the one living in a magic kitchen, who didn't beg for love or suffer for it, but was happy to share a limitless love with anyone who wanted it. Everyone was welcome in my kitchen, and it was there that they found the encouragement they needed to love themselves.

Dhara had begun to see the world as it was. The long journey from the Plaza of Hell had been emotionally and physically arduous. There had been many ceremonies like the one Miguel created for her that morning, opening minds and hearts. There had been many tears, revelations, and happy celebrations. As the groups finished their walk up the Avenue of the Dead, she sent Miguel's apprentices away. It had been a good day, but there was more for her to do, and she must do it alone. She felt Miguel nearby, giving support to her dreaming, but she couldn't see him. It was late in the afternoon, and he'd been missing for hours. People were descending the Pyramid of the Moon, meandering back to a nearby village, and the few students who had been following her wandered off to join the mingling crowd. There was a stone platform at the base of the pyramid; it was hot, sending a searing comfort through her body as she lay upon it, facing the sky. What was it she wanted, and from what power? Why did she still feel this urgency to finish something, to find the final piece in this unnamed puzzle? Had she ever wanted more than God's love, and had that love ever been in doubt? Perhaps there was doubt, and even traces of fear, but she was eager for battle and prepared to win. She took a deep breath and exhaled loudly. She opened her heart to the heavens. She emptied her mind, and she waited.

Without warning, everything that had happened until that day was made inconsequential. Her birth, her life, her struggle for understanding—all of it rushed from her like a silent, silty river. She was lucid, feeling life's power surging through her, but the familiar currents of memory were gone. She felt flimsy, weightless. She was not where she had imagined herself to be when she lay down a few minutes before. There was nothing warm beneath her, and no vaulted sky restrained her. She was nowhere and she was everywhere. Her breath came easily, and it seemed to fill every vacant space in the universe. She felt the urge to smile, but it wasn't just her face that registered the pleasure; it was everything known to her. Things, people, and places lit up with joy. She couldn't name things, or comprehend things, and yet, clearly, it was all her. She gasped with intense delight, her feelings balanced and complete at last; but just as this sensation of lightness had come swiftly, it was swiftly ripped from her.

In an instant, the weight of her own beliefs began to crush her. The noise in her head grew deafening. The *mitote,* that war of words within the human mind, the thing that Miguel had turned into a familiar mythology, was suddenly real. It seemed that every human on the planet was yelling at her or at someone else. Everybody was shouting, nagging, arguing against the truth, and their noise was unbearable. There was anger and raw fear behind the noise, and the intensity was shocking. Even more shocking was the realization that every voice was hers. Every argument, assumption, and conclusion was a part of her own thought process; every judgment came from her. Every complaint and contradiction was a reflection of her. She was the mayhem, the deafening noise in her own head. She was the liar, the deranged storyteller. She had imagined herself to be an angel of life, but death was closing in now and dulling her senses. All she could hear were messages of fear. Terror seared her brain until, suddenly unleashed, it thundered through the ruins of Teotihuacan.

It seemed she was fear itself, waging a merciless battle against the fragile body she inhabited. It was an impossible, horrible sensation. In one stabbing instant she saw that it was she who was the defender of fear and the unmistakable voice of knowledge. She was the tyrant that occupied this human being, and she was the way to salvation.

Dhara screamed, then screamed again. Her fear was amplified by every stone and tree as her screams came back to her with increasing force. Taking no notice of the world around her, she cried out yet again in agony. She let unbearable grief swell within her, and felt her heart break from it. She felt the ravaging pain until it was spent, and fear began to loosen its hold on her. Was this the final abuse of truth? Was this the last judgment, the end of believing? Her life had been built on countless falsehoods that had accumulated over a lifetime into a billowing fog of misunderstanding and distortion. She had asked for comprehension, and she now knew what the answer felt like—to the tyrant, to the lie.

Truth had no mercy, Miguel had said. He had told her this would happen, but how could she have imagined the pain? How was she to recognize herself after this? What would she do now, seeing everything and believing nothing? The questions continued, until they, too, lost power. As her cries weakened, she began to breathe in a slower rhythm, and in time she opened her eyes.

Standing above her was a beautiful woman, her lovely features distorted in sorrow. The woman was tall and elegant, and she seemed to command the wind, for her red hair tossed and curled like leaves in an autumn storm. Dhara imagined she was seeing herself, until she blinked away the tears and saw the shape of the pyramid behind the woman, and through her. The woman wasn't there, but her anger could be felt. She wasn't there, but her pleas could be heard.

I am you, the vision said. *We are each other.* Dhara blinked again. The image was fading. As it disappeared, the lowering sun illuminated the Pyramid of the Moon, paying tribute to all life's reflections. *Don't leave me! Don't leave me!* Dhara heard herself cry in a voice she no longer recognized, and then all was silence. There were no more tears left in her, no more agony. She lay back upon the cooling stone and let sleep wash over her body, the body that had supported her in her warrior's work.

GRANDFATHER! YOU FRIGHTENED ME!"

Sarita was sitting on a low wall at the base of the Pyramid of the Moon in Teotihuacan, as the afternoon sun wavered just above the far hills. It was nice to be here again, she thought. The moment seemed of little consequence to her son's life, but it felt right to be here, even in these circumstances.

"I frighten myself," don Eziquio retorted, wiping dirt from his leather pants. He had leaped onto the wall to be beside her, and now he looked around to view the scene. His black eyes squinted under the wide-brimmed sombrero. "Where is everyone? Where are all the characters in this peculiar play?"

"They come and go, without explanation," she said. "It's as if I were dreaming."

"Are you sure you are the one dreaming, *m'ija*?

She shrugged off the question. "It seems I've been sitting here on this wall forever, watching my dear Dhara. See there?" They could see Dhara on the platform, lying on her back and whimpering softly. The sound reminded her of a child's bereavement over a lost toy.

"Why does the woman weep?" Eziquio asked.

"She weeps with relief. Her pain is finished, Grandfather."

"Pain?"

"She was distressed, undone. Miguel would call it the natural result of—"

"An encounter with truth," the old man finished.

"She has been brave, faced herself, and—"

"And the fortress is cracking!" The old man settled himself next to his granddaughter, intrigued. "This is the part I relish—when the mind surrenders."

"Grandfather," Sarita scolded. "She was devastated!"

"And rightly so!"

"She needs help."

"Your boy needs help, remember? His heart must beat on its own again."

Sarita looked at the old man, bewildered. "What do you imagine we are doing? By seeking out the memories—"

"We must go to the heart, girl! Go to the heart!" Eziquio leaped up again, jumping off the little wall and onto the ground, running lightly toward Dhara to offer his favorite kind of mischief. Sarita felt an impulse to stop him, but sighed instead, marveling at his agility and speed—remarkable, for a man so long dead. Dhara had once held the key to her son's heart, Sarita acknowledged. Miguel's heart . . . perhaps her grandfather had more than foolish tricks up his sleeve.

As she watched him sprint toward the stone platform, her attention went back to Dhara's plight. Her son had pushed her too far, perhaps. He had challenged her too soon. Every good warrior craves battle, but rarely are they ready for it. This world was about knowing things, as Lala loved to remind her. Miguel was introducing his apprentices to a world of not knowing, of never knowing. In this other world, surrender was everything. They would each experience a new kind of grief before they could find equilibrium there.

As she contemplated these things, Sarita felt a different presence close by. She turned to see Miguel, the very son she was thinking about, sitting beside her. He was the Miguel she had last seen in the real world, the knowing world, wearing his silly hospital gown with a spattering of blood at the hem.

"My angel!" his mother gasped. "Are you well? Have you come back to us?"

"I'm with you now, Sarita. Who knows about the rest." He looked toward the stone platform and Dhara's forlorn figure. "I remember this," he said.

"You showed no mercy, son."

"She asked for none."

"Did you not think to give her comfort?"

"That day, she wanted none," he said. "Of course, she got it anyway. Her mind didn't break, *Madre*. What you see is the fear of breaking."

"Still, it is painful to watch."

"A few more tears, total surrender, and the universe will open to her. Awareness will come rushing in, begging to be embraced."

"I wonder now . . . must the mind be so humbled for that to happen?"

"Must the liar be silenced?" Miguel asked. "Must the tyrant be overthrown?"

"The mind," Sarita said with slow deliberation, "is making

this excursion of ours possible, Miguel. Without the memories, both yours and mine, there would not be this." Her arms swept in a wide circle, indicating the whole of Teotihuacan and the moment they both occupied. Her son nodded, looking at her with interest.

"Dhara was being the best kind of warrior on that day," he noted. "You, on the other hand, have avoided the real conflict."

"Oh, and why—?"

"I can see you humoring knowledge, the temptress. I hear it in your words and see it in your actions. You are indulging hope, the disabler. Most importantly, you are delaying the inevitable."

"Which is?"

He looked deep into her eyes, until she turned away.

"You must let go of Miguel, *Madre*."

Sarita frowned at her son, then drew her bag closer to her. She had come a long way since she'd found him in the eternal tree. She was close to having all the memories she needed to rebuild a structure and an identity. She was too close to risk failure.

"I must do nothing. I am delaying nothing. Don't speak to me as if I were your mother, worrying that you'd stayed out too late. *I am not your mother!*" Suddenly, without warning, her eyes flashed and her voice boomed. "I am not a thing to toy with! I will fight against the very elements themselves, if I must— until Sarita's world is whole again!"

In that moment of rising fury, the wind picked up around them; and Dhara's distant cries echoed against the rock walls again, as if expressing the sum of all human rage. Finding no corresponding response, the rage eventually subsided. The moment calmed; the fire in the old woman's eyes dimmed to a haunting sadness. She tried not to look at her son. The wind had softened, but thunder drummed from the crimson horizon. Birds scattered, tweeting nervously as they circled and finally settled in the low-lying shrubs.

And then, as if nothing had happened, all was at peace. Mother and son sat in silence. In the back of Sarita's mind rose the question of what would come next. She was no longer certain of her actions. Where was don Eziquio, and what was left for her family to do?

In the distance, Dhara's soft whimpers echoed against the ruins and lifted into the dusky sky as she fought against surrender, and lost again. In time, as the sun nestled comfortably among the western hills, pain dissolved into silent awareness. . . .

The day was over. Dhara had fallen into a dream. She could hear the village dogs barking in the distance and the whistle of a hawk overhead; but in her mind, she had left this sacred place in the high plateau of Mexico. She was now somewhere else, someone else. There was an ocean heaving below her, its waves roaring against rock and its cold spray biting at her bare toes. She was aware of her body, naked and wet, hanging from a cliff's edge, her bloodied fingers digging into the cracks and fissures of a rock. The sea was eager to swallow her whole; there was no one to save her.

She tried shouting out a warning to all endangered souls, but no sound emerged. She tried to think, and could not. She gasped for air, but no breath came. She was losing strength and had no will to fight. She felt the fires of terror burn through her body; her courage was gone. It seemed that the sea would take her now, and she could imagine nothing more terrible. All the torments of human existence, all the pain and the insanity, seemed preferable to this—succumbing to the will of a depthless ocean.

"Truly?" asked don Eziquio, who had been watching from a perch on the rocks above her. "I give you a chance to jump, and now you say that any sort of human torment would be better?"

Hearing his words, Dhara caught her breath. She rallied her strength, and the roaring waves stopped clawing, clamoring, beneath her.

"Who are you?" she gasped.

"What possible difference could that make?" he answered.

"Are you God?"

"If you like."

"What should I do?" she asked hoarsely. "Should I let go? Keep fighting?"

"To whom do you refer?"

"I'm sorry . . . what?"

"Do you speak of the one clinging to solid rock, or the one clinging to the self? Do you refer to the one who is stupid or the one who wishes to be happy? Which is it?"

Gulping for air, Dhara felt as if she were already drowning. "What?" she repeated.

"Are you happy, or are you stupid?" challenged the old man.

"That's not a choice!"

"Really? Are you sure?"

"It's not; it's really not," she said, her voice choked in fear.

"It's the *only* choice," he barked. Then, as his booted feet pushed off the cliff, he soared past her head. In that one gleeful bound, Eziquio took flight and disappeared into the sea.

"God!" gasped Dhara as she came awake. "God!"

She was still lying on the rough stone of Teotihuacan, alone. How long had she been there? Daylight had faded from the sky. Even in summer months, a chill swept the high plains in the evenings. It seemed cold already. Shivering, she stood up slowly. The ruins of Teo had turned to a circle of crouching shadows, and there was no one left to walk down its darkened avenue beside her.

She remembered her ocean dream, remembered her encounter with the old man. She had imagined that being face to face with God would be different. She had imagined that he would look like the Indian spiritual master Sai Baba and speak in a voice like storm winds moving through lotus flowers. Since childhood, she had imagined that God had a job for her. He wanted her to speak to humanity, to be a messenger, sharing her wisdom with the world. Hanging from that cliff, her existence imperiled, she had anticipated his divine intervention. He would save her from certain death, she'd been sure, and direct her toward her true purpose. Instead, he had taunted her. Instead, he had left her with the vision of an imp flying fearlessly into the vast ocean. She was left alone, with only that.

A dog barked. Then several others joined in, their voices sounding like the howls of old women. Life was mocking her, it seemed. Well, she was a thing to laugh at, wasn't she? She, who had once been so certain about so many things, was now clueless. She, who had raged against the teacher and battled her own beliefs, was left with a choiceless choice . . . to let herself go and fall into mystery. Mystery. That simple idea seemed to melt the fear in her. When she dared to take a deep breath, the shivering stopped. She felt better somehow. Amazingly, she felt calm. She wanted nothing. The old certainties held no temptation. Let humanity believe what it would—she was free from those concerns. For one moment, she had stepped out of the heavy fog, broken a thousand heavy chains, and was wondrously happy.

W hat have you done?" Sarita asked accusingly. "Don Eziquio, this is too much! Really, really, it is too much!"

She and her grandfather watched as Dhara walked wearily down the avenue to find some supper and a bed. Night was closing in, and the waning moon, floating above them like a beggar's bowl, was the only thing brightening the sky.

"Not at all. This is typical," Eziquio said. "It happens every time. 'Show me no mercy!' they always say, and afterward they plead for it. They refuse to fly. They moan, they cry, they blame and complain; and sometimes their heart stops. That happened once when I was teaching. I was in Sonora, working with three apprentices, and one of them collapsed where he—"

"You were playing mean tricks on Dhara!"

"Tricks!" he shouted defensively. "My so-called tricks are transformational! Miguel's too, I might add. Think about it: Did my grandson play tricks on this woman, or did he save her from terminal stupidity? Look at her now! At peace, and carrying a wisdom that all the sages in the world could not have provided her!"

"Still, change is difficult."

"She *wanted* change, my girl. She pleaded for it, and now it is done. All is right with the world, Sarita, for Dhara is learning to save herself."

"She was determined," Sarita conceded.

"She was zealous, as so many are. Like the rest, she could not stop searching, demanding, ripping her heart open to see its contents. With truth come pain, screams, release—like the sweet agony of physical love."

"She was certainly devoted," Sarita said, nodding.

"Indeed, and her devotions took her to . . . India?"

"India! Yes!" said Sarita, remembering. "We all went to India in our turn. I met him there—Šri Sai Baba, a divine being who walked this planet with great humility. He was . . . well, he was . . ."

"God," offered the old man, winking. "Is that not what you said? That he was God?"

Sarita shrugged. "We search for God everywhere," she said uneasily. "That is natural."

"There is one place that foolish people never search." With a sweep of his articulate hands, don Eziquio indicated his own body, top to toe, bringing her attention to the human that he was, or once had been. Shaking his head, he took a deep breath of night air and let the sounds of summer evenings fill the silence. "The mind is a panicky thing," he observed, "much like a boat with missing planks, bobbing desperately over an ocean of tranquillity and refusing to sink into it."

His granddaughter was silent. She was thinking of miracles, of the mysterious forces of life that seek willing allies. Like everyone, Dhara wanted to be such an ally; like anyone, she feared the cost would be too high. Sarita, too, had often clung to convenient stories and familiar ways, fearful and unsure. She, like everyone, preferred to know who she was and where she was going. Who would eagerly give up the safety of familiar knowledge for the terrors of obliteration?

"How can infinite life be obliterated?" her grandfather asked, following her thoughts. "Are we life, that ocean that swells eternally, or are we little broken boats?"

Broken boats or determined swimmers, Sarita thought, we will fight to survive in ways that we understand. Why did Miguel demand such courage from people? Did he imagine that surrender was easy? Had he not also faced the ocean, so deep and so frightening, and made his own choice? Of course he had. He, too, had abandoned the lies. He had died while he still lived, and he had done it alone. He knew the difficulties.

"He knew, yes," echoed Eziquio, "and he persisted. Even you, the teacher, would not have pushed him out of the boat."

"I was also his mother, remember—just as Miguel was the wise master but also Dhara's husband." Well, perhaps not her husband, she corrected herself, but the intimacy between them made surrender more difficult, not less so. The woman had been seeking a father and healer, redeemer and savior. Instead, her reality had been dismantled and undone. This did not feel like salvation to her; it felt like sinking, slipping down—but where, and to what?

"To absolute awareness." The old man laughed, then added, "What a repugnant fate!"

"Awareness does not replace the small comforts of life and the warm refuge of our beliefs," Sarita argued, knowing, even as she spoke, that it was a frail argument.

"In my day," Eziquio said, "it was considered a warrior's greatest challenge to rise beyond his own beliefs. Now it is seen as a tactless assault against one's identity. In my day, we would chase anger to its borders, cutting through thorn and weed until we found our way to the open fields of awareness."

Sarita said nothing, remembering her own journey. The old man plunked himself down on an outcropping of rock and tried to catch his breath. He removed his hat, wiping his brow with the back of his hand. Setting the hat on his lap, he ran a trembling finger along its brim.

"I have come to help in this quest of yours, Sarita," he resumed. "To finish it, we must dare to look into the heart of things, and encourage truth to conquer the lies." He paused. "You do see the lies, no?"

"The biggest lie is that there is something within us to fear."
Sarita sympathized, looking at the woman in the distance.

"A good teaching to remember," the old man suggested, "should you ever wish to become a wise old woman."

"That, my dearest, is not amusing."

"It is amusing," he said, smiling. Standing up again, he executed a short and lively dance. They both broke into giggles, their laughter splashing into the evening air like ocean waves. "Everything is so very funny, my little angel, if only the living could see it. In my day—"

"Oh, enough!" she exclaimed in exasperation, and they laughed again. Sarita stood up wearily. Happy and tired, the two of them watched as Dhara's silhouette met the end of the ancient avenue and disappeared into darkness.

"Someday, even she will laugh," the old man said. "The living have a simple choice. They can pay a high penalty for believing their lies—sheer stupidity!—or laugh at themselves and be happy. I made my choice a century ago and have been laughing ever since . . . and, of course, I will be laughing forever after."

Sarita nodded, smiling. She was glad that the sun was setting on this memory. Sighing, she raised her hand to give don Eziquio an affectionate pat on the shoulder, but he was no longer standing beside her.

In his place was the woman who called herself La Diosa, the spectral incarnation of knowledge. She was not looking at Sarita, and spoke no words to her. As she stared in the direction Dhara had gone, her face took on the look of a hunter, measuring distances and sensing the wind. Sarita watched her, wondering. What did she want; whom was she hunting at this moment? The creature's eyes said nothing. With an unnatural calm, La Diosa simply watched the horizon, apparently considering the many hues of twilight that rimmed the western sky.

The mind is easily humanized, because it has distinctive appetites. Like the biological creature it occupies, the mind hunts for food. Hunting patterns are habitual; an animal learns to prowl for the same kind of food in the same kinds of ways. The human mind may be a virtual creature, but it develops similar feeding patterns. It describes itself, based on all the information available, and then it gradually develops survival strategies. Survival in this case means a sense of itself as real—as real as the matter it occupies. Forming an identity, or the main character of the mind's story, is a start. But that true sense of hot-blooded "realness" is strengthened by the mind's interaction with human emotions. It doesn't take long for a child's mind to recognize how a word, that tiny expression of knowledge, stimulates an emotional response. Words soon become tools for the hunt, and emotional feeding patterns are quick to follow.

Knowledge is the main character of my story, and everyone's story. As I reflect on Lala, the huntress who is blind to her own nature, I feel only compassion. She created the story of Miguel long ago and wants nothing more than to keep telling it. It's been a long time since I felt the need to be me. At this point, it is impossible even to remember such a time. Completely free from the pull of knowledge, I can imagine the main character of this shifting story any way I wish. Throughout my lifetime, I've been aware that Miguel meant different things to everyone; but now, from a more detached perspective, I see that Miguel is simply knowledge—more precisely, my knowledge. The way I accumulated knowledge over a lifetime, harvesting beliefs from the seeds of opinions, was unique to me. The voice of knowledge has a stressful effect on the body, as most people understand. Once I heard it, once my own voice became apparent, I was compelled to investigate it. Recognizing it as me, I was compelled to change my mental habits and appetites. I chose to develop a taste for truth. With truth, I realized, I could set this human free.

Achieving personal freedom means liberating ourselves from the penalties of knowledge. Change one thing—one idea, one habit—and the frontiers open. Freedom is a beloved idea in the human dream, an idea that is not fully comprehended. Freedom begins when each individual mind dares to liberate itself from the prison it created. We are free when the war in our heads is over. We were born authentic, responding directly to life, but we lost our authenticity on the battlefield of ideas. We can be authentic again, of course. We can become immune to the effects of knowledge. We no longer have to believe ourselves, but we can learn so much from listening—listening to our own thoughts, to the opinions of others, and refusing to pay the emotional price they demand. We can see, without making assumptions. Change one idea, make a different agreement, and the prison bars begin to bend.

Human knowledge has proven its worth and potential. Knowledge gives the mind a landscape, a virtual space where it can use words to explore and play. As words make objects appear in our imagination, this landscape becomes richer with visions and possibilities. Knowledge is the backdrop to all human dreaming. It's meant to serve as an ally to awareness. From that perspective, we can also see that knowledge is the angel that delivers us . . . or the demon that possesses us. That recognition gives us the power to take control of our stories so that they more consistently reflect the truth. Taking control of the story is our first tentative leap to freedom.

It's simple to see that freedom, for every man and woman, lies in the mind's willingness to alter old habits. First, though, the mind has to hear itself and change the conversation. It was for that reason that I gave my followers some new "agreements" to make with themselves. Without realizing it, they had spent a lifetime making choices that did not reflect their present consciousness. For example, like most people, my students had made a pact with themselves that asserted life was unfair or that they were unlucky; they had already decided that they were victims. Having agreed to such unspoken contracts, they felt compelled to live up to them, struggling for years to react in ways that felt uncomfortable, but were accepted by society. Fear is contagious, if we agree it is. Judgments are inevitable and feeling victimized is normal . . . if we believe this to be the case. It's common practice to gossip, to be intrigued by someone's misfortune, and to get caught up in everyone's drama; but that common practice is normal only if we say so.

And we can choose *not* to say so. To help with that, I suggested some new agreements—what I would come to call "The Four Agreements"—to alter old behavior patterns. I told them not to take things personally, and to stop making assumptions. I asked them to be impeccable with the wondrous gift of language: the words they spoke and the words that comprised their innermost thoughts. I asked them to do their best in every effort. Four simple agreements. I also reminded them again and again not to believe, but to listen. Put into practice, these new agreements would cause a disturbance that would alter their reality.

Ultimately, I asked my apprentices to surrender—not to me, not to anyone, and not for any reason—only to surrender. The mind invents words and determines the meaning of words, and the word "surrender" means something unpleasant to most people. It means to capitulate and admit defeat. The mind can't comprehend the benefits of total surrender, but the human does. Without interference from the mind, the human moves toward surrender as a caged animal moves toward freedom. It knows how to give in to the demands of hunger and the need for sex. It knows how to fall into sleep or into love. In satisfying those physical demands, the body is renewed and strengthened. Surrendering to life is an act of power. Giving up on old stories is an expression of gratitude—from the mind, to a loyal body.

I surrendered long ago. My war with knowledge is over. I don't think anymore. I see; I listen; I respond. Words are not the most compelling thing—not for me, not for anyone. We are far more attracted to the force of love. And yet few people can imagine how many old agreements have to be broken for this kind of force to prevail.

In the aftermath of my heart attack, I must surrender again, giving up every expectation and every assumption. I must give this body over to life, saying yes to all its blessings and its disappointments.

If this body survives, I will start again, meeting each day with a laugh . . . and, of course, I will be laughing forever after.

W HY *AGREEMENTS?*" LALA ASKED WITH CON-
sternation. "Why not commandments? Why not
the New Commandments! The Four Sacred Laws!
The Pious Promises!"

"They can break an agreement," don Leonardo sighed, "or
change it. There is no judgment or retribution."

"How does that keep humans in check?"

"Have commandments kept anyone in check? Are vows
never broken?"

"Vows are broken, and then confessions are spoken," she said.
"They commit a sin and then judge themselves. God judges them
and they judge each other. They must be thoroughly judged."

"To judge, and to find yourself guilty, is to go against
yourself."

"Oh, of course. The Law of Impeccability." She yawned.

"The first *agreement,* my dear. If you dishonor an agreement, learn from the experience and do your best next time. It is an act of grace to forgive. It is an act of love to offer yourself another chance."

"It is a godly act to punish the backslider."

"Yes, I see how that has been the strategy—to turn browbeaten children into guilty men and women—but has humanity lived so happily under these laws, *señora*?"

"Like you, happiness is pure fantasy."

"Like me, *you* are pure fantasy."

They paused. Sarita was absent, but they could still feel her disapproval at their bickering, so they held off. With this pause came an awareness of their surroundings. It appeared that they were watching a private ceremony on the top of the Pyramid of the Sun.

"This is an interesting event," don Leonardo commented, shifting his focus.

"Where is this?" asked Lala with alarm.

"Quite obviously, we are on the great pyramid. I never had the pleasure of visiting this place."

"Nor I," she echoed quietly.

"Aha! You have been promoted!"

"How is it possible?" asked Lala, catching her breath. She was uneasy, in spite of herself. The Pyramid of the Sun represented life, not its reflections.

"Never mind how," he said, stepping closer to the scene. "Let us give it our full attention. I recognize Miguel, my grandson. He is not changed much since his student days."

"Your decrepit daughter should be seeing this."

"My venerable daughter, *señora*. Please show some respect," he chastised. She had a point, though, he admitted to himself: Sarita should be seeing this. "You and I have been entrusted with this event. We are here in this glorious spot, where human intent has connected sun to Earth. I see Miguel, smiling and silent. I see a woman with him . . . but her face is unfamiliar to me."

"The bookmaker."

"The bookmaker?" he asked, surprised.

"Your grandson's first book is being printed, and it will be read by people of all cultures. It may just change the world, sir—through language!"

"Wonderful!" exclaimed Leonardo. "I did not live long enough to see this!"

"Indeed, there is nothing so wonderful as the printed word," she said, picking up a few lost shards of enthusiasm. "Do people ever doubt words that are put to papyrus?"

"Paper. No, not often." He took a deep breath and nodded his head approvingly. "*The Four Agreements*. A bold perspective. Clean. Simple. And yet disruptive . . . compelling."

"The dream is fine the way it is."

"Sickness is fine?" don Leonardo snapped. "Tyranny is fine?" The woman was testing his patience. "Are you saying that fear and retribution are good companions, or that violent responses lead to pleasant results?"

"It appears you are taking the subject a bit personally."

The old gentleman looked into her fiery eyes and sensed the danger there. It served no purpose to encourage this one, he reminded himself. Words incited conflict, her favorite food. Be they sweet words or bitter words, she would feast on the result.

"I have great respect for your powers, my dear lady," he said sanguinely. "I wish only to emphasize that you could apply them to better effect."

Lala stared at him, unable to think of an appropriate response.

"The bookmaker seems well-meaning," Leonardo observed, "but what is her precise intention?"

"She is reciting prayers, obviously. She is performing a cere-mony of her own design. She hopes to conjure success through heartfelt wishes."

"She stands beside a *nagual* man, and presumes to wish," don Leonardo commented, then smiled. Just the sound of that word enlivened him. *Nagual.* Total power. The *tonal,* in his cul-ture, referred to matter. Beyond matter was sheer mystery, the limitless and the unknowable, that which was impervious to knowledge. A *nagual* man was one who knew himself as in-finite potential, as the force of life itself. Don Leonardo walked toward his grandson and reached out to touch him. There was no one there to touch, and nothing to feel but the power that still lingered within a gossamer memory. Wait! There it was! As the old man's hand felt the air and his fingers trembled, his smile grew into quiet laughter. "Intent is the physical force of life," he said. "Intent runs through this man."

Lala turned back to watch him, and curiosity took hold. Her hand began to imitate his, moving through the air, around and between the two people who stood, windblown, at the summit of the pyramid. "Intent, you say, or intention?"

"Intention is the mind's work, *señora.* As you put it, a wish and a prayer . . ."

". . . and a hope," she murmured.

"It takes something beyond hope to bring a dream to life," Leonardo said. "It takes action—action that is fueled by faith in oneself."

"Faith in oneself?" said Lala. "You are the father of blas-phemy, sir!"

"So says the mother of lies, madam!" the old gentleman shot back. Satisfied that he'd had the last word, Leonardo shifted his attention. In that instant, they both evanesced into sun-shine, allowing the dream to rearrange itself and move on.

Father of blasphemy, mother of lies. How silly words can sound when we forget their true purpose. We are all tempted to accuse, and easily persuaded to defend a favored illusion, an idea of ourselves that only words can explain. By investing all our faith in those words, we become the illusion. We are knowledge, struggling tirelessly to find the words that best describe our journey back to truth.

Nagual is a word I grew up hearing, a word that caught my imagination early on. Within my family, *tonal* and *nagual* were familiar ways to describe the totality of life—matter and pure energy. My grandfather Leonardo loved to tell me about the Toltec ways and traditions, and he shared his understanding of these things with joyful enthusiasm. I can feel his delight even now, as I recall how he helped me rekindle my own love for life. Don Leonardo told me many wonderful stories and guided me toward mysteries that couldn't be *told* but might be *experienced*. As a teacher and guide, I have also told stories in order to excite wonder and a deeper curiosity. One, in particular, conveys an essential lesson in awareness. I have told it many ways, but the message is always the same. . . .

There once was a man who, like many humans, became aware of himself as the infinite force of life. This happened to him in a sudden moment of inspiration. This kind of inspiration can happen to anyone, at any time. In this case, the man stood under the stars one clear and silent night and was captivated by what he saw. This happens to all of us, this sudden and strong appreciation for the majesty of the universe; we are suddenly amazed to see beauty everywhere. We take on the eyes of an artist, and beauty is *all* we see.

So . . . in that moment, the man I'm talking about understood everything—*everything*—without words. It didn't matter how long ago the stars had sent information across the landscape of infinity, or if those stars still existed. He was receiving their messages in that moment.

Having looked at the night sky, we all know that the darkness stretching between stars looks like empty space. We may also know that this space is much, much bigger than the space that is occupied by all the stars put together. More than two thousand years ago, in the great civilization of Teotihuacan, the Toltec people referred to the space between objects as the *nagual*.

Let's say that this man, standing under the stars that brilliant night, suddenly looked down at his hands. There, too, he saw the universe. He saw that his hands were made of millions of atoms, in the same way that the universe was made of stars. Like the stars, the atoms in his body represented the *tonal,* or manifest life. He then realized, without any doubt, that the *nagual* creates the *tonal.* He could see that the light-filled emptiness was responsible for the creation of all matter. The *nagual* was total power, the infinite force of creation. The story goes on to talk about the man's excitement at his discovery, and his desire never to forget its meaning. He knew that his experience could easily be forgotten in the distraction of human existence.

So what are we? Are we the *tonal,* or are we the *nagual*? Are we matter, or are we life? Humans have been asking this in different ways for thousands of years, without realizing the simplicity of the truth. The truth is life and death, a simple binary formula whose mathematical symbols are 0 and 1. In the language of science, this means energy and matter. In religious storytelling, it is God . . . and creation.

The stories we tell about truth often lead us further into distortions and deeper into our own fears. We don't need to prove that life exists—if we tried to do so, we would be doubting our own existence. We are alive, so life exists. Death, or matter, obviously exists as well—everything that has been created also has an end. Incarnation is the process by which life creates matter, moves matter, and becomes matter. The man in my story knew he was the *nagual,* the force that animated his physical body. The body, or *tonal,* was his creation. The body was his sanctuary, a place he loved and respected unconditionally. Every place he occupied deserved the same respect. This planet, an object among billions of objects in the vast landscape of life, was also his home, deserving of his respect and love.

A *nagual* man knows himself to be the force that creates existence and moves matter. He sees that all else is temporal. All else—like thought, like words—is a mirror distortion. To know these things is to know truth.

Is it not true, my angel of love, that walking hand in hand into the dream of life . . . every step is blessed by God?

The wedding ceremony was being held at the home of an apprentice in New Mexico, surrounded by red cliffs and embraced by sapphire skies. This was a good place, an enchanted place. Sarita could feel the hum of Earth-music beneath her feet, and her ears heard sacraments in simple birdsong. The scent of life was evident in desert blossoms and piñon trees, as the summer sun scorched the air to a crystalline brilliance.

Miguel's words to the gathering were strong, and so was his intent. Listening as a remote observer, Sarita marveled at the change in her son. She had forgotten how evident his power was at that stage of his life. Back from a journey to Hawaii, he was quickly changing. He had found someone who was eager to publish his first book. He had fallen in love again and begun planning for a new family. He had cut his long hair, altered his style of dress, and started jogging every day. He was still young at forty-five, and even more handsome than his father at that age.

"José Luis," she whispered wistfully. It was good to think of her husband again. At the time of this wedding, he was so recently gone, and his absence had changed her in many ways. After their trip to India, José Luis had begun to show symptoms of fatigue, but he had dismissed them as trivial. He had refused to seek help, and she'd lost him before anyone understood the seriousness of his illness. His death had been unexpected and devastating. Had that loss also changed Miguel? It seemed that *something* had. He began to tell a new story, another kind of story, with a different cast of characters. Dhara had moved on, and he was inviting another woman to share his life. He was getting married. A few friends and students were here to witness the ceremony; there were no family members present. He recited the words happily, as if he were a boy again, eager for the next adventure and indifferent to its consequences.

Is it not true, my angel of life, that in the eternity of my joy, the smile on your face reflects the love in my eyes?

He was igniting a new flame, yes, but he was also extinguishing an old one. In spite of Miguel's almost childlike enthusiasm, this ceremony was not being undertaken with the recklessness of a boy, but rather with the deliberate calculations of a man. By committing to this woman, he would be ending old dreams and welcoming new ones. It was strange to see the wedding ceremony from this perspective—one more moment in his life that he hadn't shared with her, and one more aspect of him that she hadn't known. She was gratified to witness it now, and thankful that he wanted her to. Change was not a challenge for a man like him. He could adapt, and he could love, no matter the person or the circumstance.

Sarita wiped a tear from her eye and watched it float into sunlight. She hadn't realized how these memories would stir her. Someone touched her shoulder, and she became aware of the older Miguel, standing at her side. He was wearing his hospital gown, of course, looking small and fragile, nothing like the man in the sunlit ceremony; still, a smile lit his face as he observed the scene playing out before them.

"A mother's happiest dream," he said cheerfully.

"You followed me here," she said, nodding. "Thank you."

"I wanted to enjoy this moment with you."

"You've looked better," she noted wryly, pointing toward the groom as evidence.

"I have indeed." He watched as the minister read the service. "Isn't it good to see your son happily married at last?" he asked, smiling ironically.

"You were happy; I see that," she said, her eyes gleaming. "You remind me of your father, *m'ijo*."

"My father made his marriage a success."

"He would never have dared to leave me—that is certain." Miguel was not his father, she mused. Strong as women might be, he was always more so. "She was a delightful girl, as I remember her," Sarita offered.

"She still is, but it didn't work."

Sarita was about to answer when she saw a bit of movement above them, on a nearby ridge. Thinking it was a deer, a symbol of grace and love, she pointed.

"Is that—?" she began, and then recognized the silhouette of her grandfather, so spry and yet so long deceased. "Is that don Eziquio?" she marveled. "He'll kill himself jumping around like that."

"He insists on blessing the occasion." Miguel smiled, then said, "It was blessed, in truth. Everyone had a magic glow that day, don't you think?"

"You seem very much in love," she agreed.

"Of course I was. I adored her."

"You say that about all of them," she said affectionately.

"I *mean* that about all of them." Miguel shrugged.

"They can accept only the love they think they deserve, *m'ijo* . . . not a limitless love, not one that sees beyond the woman, to the truth." She smiled, her face touched by emotion. "Still, it seemed so right between you and this girl."

"There was no right. There was no wrong." He put a frail arm around his mother. "I tried. I failed. That's it."

"Son, look at your joy, your fearless resolve! The bride is radiant, and you have such excitement in your eyes!"

"I made a big effort, and so did she," he said, watching. "The desire was there, the sex was wonderful—"

"Miguel!"

"—but should it have been such an effort? We rarely agreed about anything. It became clear that I embarrassed her. She was reluctant to introduce me to her friends. When she did, I felt like a squid in a pool of exotic fish."

Sarita had to laugh. "Such a brilliant and exceptional squid!"

"This was not a match," he insisted, indicating the couple getting married. "It wouldn't have lasted, but it could be argued that you and Dhara didn't mourn its collapse."

"Did you expect otherwise?"

Did he? Miguel frowned slightly, wondering. "You both wanted to drive Miguel's dream, *Madre,* and you could not," he said. "It was mine to live."

Sarita looked at him, considering the vehicle he would need to navigate through a human dream once again. His body was compromised, but there was still a great *nagual* man at its helm. His had been a remarkable existence. His life had been a monument to loving, and his body was its instrument. He would come back and give more. She would see to that.

"How long did this one last, in the end?"

"Three months. Three months of discords and perfect harmonies."

"And then what? What should I expect to see next?"

Her son turned to her, looked her directly in the eye, and said, "Are you ready?" It was a question he always asked apprentices who insisted on seeing things they were not remotely prepared to see. In fact, he would always ask them three times. Realizing this, the wiser ones would stop demanding an act of power from him. Those who didn't were warned, but only barely. "Are you ready?" he would say a third time, and then reality would shift before they could answer. Visions came, thinking stopped, and the truth roared in like a punch to the stomach.

Sarita had no warning at all.

"Are you ready?" he said softly, weaving the words under the jabber of piñon jays and the rush of hummingbird wings . . . and then she was standing in an unfamiliar room.

It was a woman's bedroom. It must have been nighttime, because a bedside lamp was on, and the apricot curtains and bedspreads had turned to flaming gold. Dhara was sitting on her bed, in her own home in San Diego. It had been a while since Sarita had visited the house, but she recalled that this is where Miguel had come when he left his new wife a few months after the wedding. He had gone to Dhara's house. He had flown back from a Circle of Fire weekend, a yearly gathering of his students and family members, and announced to everyone there that he was releasing his bride from her vows. He had left his car in Tahoe and flown back . . . to Dhara.

Sarita could see Miguel, as he was then, sitting in an armchair several feet from the bed. His eyes were closed, and he said nothing. He had arrived late at night. Dhara's son had let him into the house. "I'm here to see your mother," Miguel said, walking unannounced down the hall to her bedroom. Dhara was astonished to see him there, but his expression told her everything. She kept her silence. An hour later, they were still sitting on different sides of the room, in silence, offering each other grace and forgiveness without commentary. There were no more battles to fight.

Sarita recalled the next weeks, as Dhara rose like a marvelous eagle, lifting Miguel away from his disappointment over the failed marriage. They traveled to Italy, and the journey shifted and changed him. He would let go of his sadness and rebound. They dined in Venice, toured the Vatican, and walked the ruins of Rome. Italy was a feast of pleasures and amusements—something Miguel could not say about his experiences in India, where the two had traveled together years before. India had not been the heady spiritual adventure for him that it had been for Dhara, or for Sarita. The place did not suit him, it seemed; but being in Italy provided the distraction he needed at a difficult time, and the precious opportunity to decide how his energies would be redirected.

Sitting in silence with Dhara this night, his heart began to heal. The two would come together to share new moments, but without demands or conflict. Sarita sighed, closed her eyes, and dreamed with Miguel and Dhara in the lamplight. She sensed a deep tranquillity now. The anger was gone, and love burned hard and bright within them both. Truth had finally reclaimed the eternal moment. No rebellious words were left to speak . . . and for this one sweet glimpse of peace, Sarita would be forever grateful. In this place the war was won, respect ruled, and consequence had no meaning.

"Are you ready?" she heard her son ask.

"Yes," she answered, but the vision was already gone. Dhara and the lamplight were gone. Sarita was back with Miguel at the wedding. The bride and groom were now dancing with their guests, and the sun was setting on the western hills. Eziquio, the trickster, skipped lightly on the rim of a mountaintop. Magic showered over his little land of enchantment and touched all its dreaming souls.

13

J AIME COULD SEE HIS MOTHER SLEEPING IN HER
little bed. He could hear the soft, slow rhythm of her
breathing, as a dim shaft of moonlight slipped through
the open curtains and made the wall glow behind her. Sarita
had taken a long time to fall asleep, but she finally surrendered,
allowing the evening prayers to do their job.

He wondered what she had seen during the ceremonies tonight. As the drums were beating and the gourds rained sound-
drops through the house, he had glimpsed many visions, but
what of her? Where did *she* go when the others were calling the
names of ancestors and reciting the old prayers? The work she
did every night, expending precious energy as the family and
followers gathered for ceremonies, seemed to be harming her
more than it helped Miguel. She had swooned again tonight,
and this time Jaime had stopped the rituals and sent everyone
home. Miguel had lived another day. Maybe he would survive

one more, two more, but they were losing the battle against death. The doctors were gradually losing his brother, but Jaime would not lose her. He would not permit Sarita to die, even to save her youngest son.

Miguel's heart had almost failed once before, Jaime recalled. On a power journey to Hawaii years earlier, he and Dhara had taken a group of intrepid apprentices to the mountaintop on the big island. Miguel had led them into the mouth of the volcano, a trek of at least a mile. Dhara had stayed behind, unwilling to deal with the heat of the sun, the fire in the ground, and the inevitable climb back to the volcano's rim. It was Miguel's task, Sarita had decreed, to merge with Earth. In that power place, Earth and fire were symbolized by the goddess Pele. Miguel did as he was instructed, courting her with rituals and daring to enter her. Pele must have been a difficult woman to woo, because on that day she showed no mercy to her suitor. Bored and humorless, she sent Miguel away without a blessing; and halfway out of the volcano, his heart began to fail.

He felt excruciating pain in his chest. Turning pale, he began to sweat, to stumble. His students reacted in alarm. Some of the big men offered to carry him, but he laughed at the suggestion. It was Jaime's sense, as he remembered the story, that Miguel had decided to teach the biggest lesson of all. Pele had spurned him, it seemed, but he would now be able to show his apprentices how a true master faces his own death.

As moonbeams chased each other quietly around the room, Jaime smiled to himself. His little brother had always been a wizard at games, inventing new competitions—new ways to play ball, to gamble, to win at chess. It wasn't like him to be satisfied with the same activities from one weekend afternoon to the next. When the bigger boys began to master a game, he would change the rules or move the goal lines. He would invent other games with different, clever strategies. As a shaman, he did the same. Once his apprentices had captured an idea, accepted a theory, Miguel would abandon it for another one. Students were kept awake and off balance. To keep up, they had to shift, to be agile—they had to, or risk losing his attention. Over time, he became the master of mythologies, giving them constant opportunities to attach and detach—from his stories and from their own.

The new domestication required new incentives with no penalties. He was the savior his students had sought, and the parent they wished they'd had. He was a friend to many, but always the teacher and guide. There was no right or wrong this time around, and the rules everyone was asked to follow were part of a game in which everyone was the winner.

Miguel hadn't wished to frighten his spiritual children that day in the volcano. The pain was so great he could barely keep his body from crying, but he knew that they would misinterpret any tears for sorrow or dread. Dying was not something he wanted them to fear. He kept talking and he kept walking, slowly but determinedly, to the volcano's rim. The women with him cried, fearing that he could not survive the climb. Men prayed and found strength within themselves.

Dhara had no idea what had happened until the drama was over and he returned to her for help. Together that night, they calmed his heart and healed the memory of a difficult day. He had survived. He was alive, but he was forced to wonder at his own game plan. Pele had refused him, even abused him, but he would try again. He would invent different strategies. He had the faith needed to win. The goddess would eventually yield and soften, and the game would turn in his favor, as it always did.

It wasn't clear to Jaime whether those students ever understood what it meant to master death. They may not understand even now. Miguel had faced physical death with cheer then, in the volcano, just as he had this time. It was a game, after all, but one that few had the awareness to play. Maybe they had listened, and maybe some of them had learned. That day in Hawaii was a day of portent, in any case. It was a warning, and its message was clear to all of those who prayed for him this night. He was experiencing another confrontation with death now; and laugh as he might, the odds were not in his favor.

Meanwhile, Sarita suffered. Jaime had been beside her since his brother was first taken to the hospital. He watched her, wondering if she was trying, even as she slept, to shame her son into coming back—just as on those summer days of childhood, when he'd had to be dragged home to supper after a long afternoon at play.

"Where do you go on these nights, Sarita?" Jaime whispered into the moonlight. "What do you hear? What do you see?"

He closed his eyes and took a deep, calming breath. He wished to go with her, to accompany her into whatever realm she had found, and to help her achieve her purpose. She hadn't told him about her inward travels, only that she had spoken to Miguel and was being guided. *Guided* . . . by whom, and to what destination? Jaime sat up straight. He suddenly felt that someone else was there, listening to his thoughts. He felt the smallest sense of something—a guardian angel, perhaps—sitting with him at his mother's bedside.

"What does he imagine is guiding his mother? Angels and cherubs on winged stallions?" Lala wondered aloud, watching mother and son in the moonlight. She and don Leonardo had just found themselves in this room, in this quiet recollection, and were troubled to see that they'd lost their connection to Sarita. In her exhaustion, she had abandoned the clarity of the trance state for a dreamless sleep.

"He imagines an elder, like you or me, sharing the secrets of the universe with his mother while he labors on, chanting ancestral names with his cousins." Leonardo smiled sympathetically at the sight of his grandson.

"He is envious? Ha!"

"Tell him," Leonardo prodded. "Tell Jaime what he wants to know, *señora*. Is that not your vocation—telling elaborate stories to taunt the mind?"

"I will not tease him. He is a believer, a friend to knowledge, and a warrior. Look how he worries over his dear mother! Don't you suppose he would give up a dozen brothers to keep her?"

"No."

"One brother, then?"

"We will not give up on Miguel, my dear. This devoted brother will keep Sarita on the path, and she will bring her youngest boy back to the living."

"I care little whether the shaman stays or goes; remember that," Lala said crisply. "These people pray for answers. It is my advice they seek, and my pleasure to give it."

Leonardo shrugged, sensing that he must change the conversation. His attention turned to Sarita, sleeping soundly in the world of living things and urgent concerns. The cause was pointless without her presence; her son would answer to no one else. As the thought touched him, so did the sense that Miguel was hovering nearby, watching and waiting. Leonardo suddenly felt the urge to act on Sarita's behalf.

"Miguel has very little time left," he stated. "It is impractical for us to observe my daughter and do nothing."

"We have been doing many things."

"Yes, we have gathered many memories, but there is more to be done. Do not argue with me, please," he said before she could interrupt. "I fear I must insist."

Lala remained silent, watching Jaime Ruiz guard the sleeping figure of his mother. It seemed that knowledge was a road that carried humans only so far. When the prayers were done, when the hope was gone, they had only themselves . . . and the vacant serenity that stretched between two thoughts. Such a place was not for her. Such a place belonged to stalkers of mystery.

"Love is a mystery," she said, finally.

"It is a word, *señora*, as you know well."

"It is a word," she said. "It is a command. It is a torment, if I say so. And yet . . ."

"And yet . . . ?"

"Before the word was ever spoken, there was . . . something."

"Something that rules words," the old man agreed. "Something that rules all dreams, all universes."

"Even hers," she said. Frowning, she looked at Mother Sarita. "Even now."

"Even now, as she dreams without us."

"It appears we are alone, don Leonardo, toying with a fragile idea."

"Which is?"

"That a son's love may ultimately be the spark that makes a miracle."

The old man gaped at her, unaccustomed to her tone and to the face of the woman who was speaking. She had an enraptured look, a look he would have expected on a love-struck girl. Where were her thoughts, those of the storyteller? Who was behind this transformation? He glanced around the little room, squinting into the shadows, but could see no one else.

"May I suggest," he said, "that we go where her visions have taken her?"

"And if there are no visions?" she asked, wondering how such a lifeless body could conjure dreams.

"Let us give chase, shall we? You are Artemis—and I, one of your handsome, tireless dogs. Let us hunt her down and see what we might find!"

Taking one last look, Lala nodded. She would follow the old man, for he seemed driven by inspiration, something else that eluded her in this moment. With a swift look into each other's eyes, she and don Leonardo were gone, leaving an old woman asleep and swathed in moonlight . . . with a son, and the smallest sense of a guardian angel, sitting close by her side.

◉ ◉ ◉

Sarita was indeed dreaming. She was dreaming of another wedding, a huge wedding. In this particular ceremony, there were many brides. There were countless brides, although Sarita could recognize only three. She saw Maria, Miguel's first wife, accompanied by her three young sons. There was Dhara, dressed in a sari of gold silk. There was the woman who was to be his wife for a few brief months, whose face she recognized but whose name had now faded from memory. And there were more brides, hundreds of them, gathered by the steps of an immense altar, waiting excitedly for the groom to arrive. The affair was being held outdoors, but this time the weather was humid, and breezes blew warm and fragrant from the sea. The mood was happy, expectant, with an underlying romantic intensity that seemed always to arise in tropical climates. Everyone seemed restless, as if anticipating sweet nighttime pleasures and passions.

It seemed that the ceremony was indeed taking place on a tropical island; where she stood, the ocean was visible in all directions. Sarita could not see Miguel, but she felt the nearness of him. She sensed his anxiety at the prospect of so many marriages, but his fear was clearly outweighed by his enthusiasm for the honeymoon. She turned her face toward the wind and felt the pure animal force of that excitement. Had she not felt it, too, as a younger woman? Was that not the sensation of life itself? Every human yearned for euphoric love and ultimate union. Every man and woman wanted to merge with life through the body of another human being. Sarita was no different, nor were the men she had loved. Opening her senses, she was transported by the feeling, banished from the boundaries of belief for an instant and immersed in love. This, if anything, was the essence of him . . . this unconstrained spirit of desire. How deeply it affected the many women in his life, she could not determine, but this was his real power. Knowledge must capitulate to this. The mind, with all its clever calculations, could never prevail against it. This force of unconditional love was in him, and it *was* him. A hundred ardent brides stood in waiting, breathlessly proving the point.

The wedding party was enormous. The crowd, with guests that included everyone she knew, practically covered the island. Sarita saw members of her family—her sons and daughters, their children, and their children's children. Old friends mingled with new acquaintances, and brothers and sisters long since gone were drinking and laughing with the living. She saw her own father standing among the hordes of guests, and standing beside him was La Diosa. It was at that precise moment that Sarita realized she was dreaming. This was not a trance, nor was it a memory. This was the kind of dreaming that sleep produced, without cause or intention. Don Leonardo might be visible, and the serpent woman with him, but it was a senseless and rambling dream. It was a romp, a diversion of her own making; and as she came awake within it, the dream began to shift to her will.

The wedding guests vanished. The grand assemblage of brides disappeared. If there was to be a honeymoon, the groom would have to return to the living. Once Miguel was back, he could have all the lovemaking he desired, marry as many women as he could tolerate, and delight in the births of a dozen new children. Until then, there would be no talk of mating and matrimony. She would finish this, in spite of Miguel or because of him. Let the ancestors follow her or return to their dusty beds. Let saints forsake her and angels fly away; she would finish this task and bring back her son.

My mother has always played a significant part in my dreaming, both while I've slept and during my waking hours. Should I survive her, I'm sure I will continue to feel her presence and hear her words. The Miguel who occupies her imagination may not be entirely familiar to me, but she loves him without condition. As remote as he might be right now, having slipped beyond my reach and hers, she worries over him still. She dreams of the women he's known and desired, all competing with her for his attention. Attention is humanity's greatest prize, and our greatest tool for awareness; but very little requires Miguel's attention now. Life's attention will determine events and bend this dream toward a distinctly new horizon. From there, more revelations await. With my brother in attendance, Sarita will rest, wake up, and resume her efforts, as every true master must.

The wedding ceremony in 1997—not the mass wedding she was just dreaming, but the little New Mexico wedding—was an indicator of my determination to change Miguel's dream. It was my feeling at the time that a new love and a new way of life would stimulate fresh possibilities. I was taking charge of my health, my appearance, and my destiny. *The Four Agreements,* newly published, was already reshaping the dream. My approach to teaching would change soon, and my relationship to life was growing more fluid, more intimate. Sarita is right to say that I was never so aware of my personal power as during those months, and the five years that followed.

That marriage ended quickly, but love never has to end. Respect doesn't die because dreams prove incompatible. Dreams die . . . and, in dying, they make room for many more. New dreams benefit from our strengthened awareness and bring with them very different characters. Some of these may see in us what others could not, and love us with the kind of passion that never fades. The wind that blew in northern New Mexico on the day of my wedding carried a high-spirited message of change. This is just the beginning, it said. Surrender once more, it said, for intent is shifting everything. Life is discarding things and reinventing things. The messenger is transforming, along with his message. The world is listening. Life is moving at full velocity. A union is coming, it said—love filled with a passion that is destined to last.

From an early age, I considered lovemaking to be my mission as a man, both biologically and morally. In adolescence, I began to suspect that girls liked it as much as I did. If girls craved sex, and my mother was a girl, then she also craved sex. That revelation was upsetting at first, but inescapable. Using my deep respect for my mother as a point of reference, I committed myself to loving women, respecting their desire for pleasure above all else. I was a male child who grew into a man, eager to love and be loved. Life was no more complicated than that. My brothers and friends were supportive of my effort, as long as it didn't interfere with theirs. Love makes humans happy. Guilt, shame, blame do not.

Romance is the story we tell about lovemaking. Poems, candles, and music are wonderful, but sex doesn't depend on them; it doesn't change or evolve because of them. Sexuality is our essential quality. We humans comprehend truth through emotion and physical intimacy, but notice how quickly we turn against truth with stories of blame or resentment. We make the human body pay for our ideas of good and evil. I learned early in my life that the truth can be felt without a story. Love, the truth of us, transforms reality on its own.

"When did you first agree not to be victimized by love?" a student once asked me. Such a question can be asked earnestly only when you have seen the distortion and refused to be persuaded by symbols. Every human mind is under the spell of knowledge—words signifying only what we say they must. We are captivated by their power, but we are the magicians who gave them power. Spellbound and unaware, we use words to hurt ourselves and others. The words we think make us afraid. The words we speak enthrall us.

When I was in my twenties, I made an agreement with myself about the word *love*. I saw it as representing the force of life in action, creating a balance of gratitude and generosity in whatever dream I shared. I saw myself in the same way—a force of life in action. We were the same. If I was love, how could I be victimized by love?

Love, the force of truth, is too often used as an excuse to deny truth. People learn to love conditionally. They love *if* . . . if they are loved in return, or if they can control the life of another person. Humanity has practiced this distorted version of love for millennia. We seldom consider the possibility of a love without conditions or judgments, and rarely embody it. The love that most humans express and experience is the opposite of love. Like the iconic tree that mirrors life, it is a copy of truth—but it is not the truth. Like the fallen angel, it is a messenger trapped within its own lies. Like Lala, it is knowledge . . . a masterful story, told in captivating and perilous ways.

W hat do you think of Lala?"

Miguel and his mother were having a conversation on the little bed in her room at home. Jaime had returned to his family, and the two were now alone together. The curtains were pulled back, and the moon was so low in the sky that it shone in Sarita's eyes. Her youngest son was lounging comfortably against the pillows, as he used to when he was a child, just home from school and talking to her about his day. Sarita, still sleepy from her long nap, didn't question his presence there, and hardly noticed his appearance. He looked like he was heading out to the pyramids: he was wearing worn jeans and a denim shirt, his hiking boots laced up and a brown felt hat set casually upon his head. He talked softly, intimately, and his voice seemed far away.

"Tell me," he said again, "what do you think of her?"

"What?" she said distractedly. Had he mentioned Lala? How far from her present state of mind was that woman! How dim a memory the name had become!

"Your guide," Miguel said. "How does she seem to you?"

"Ah, well, the woman is exasperating," Sarita said, rubbing her eyes. "Bossy," she added. She struggled to clear her mind. "Yes, very bossy—and often ill-tempered." She was beginning to remember Lala more clearly. "And vain: she imagines herself to be quite stunning and smart!"

"Hasn't she been helpful to you?" he asked with evident concern.

"I suppose that remains to be seen. She is brimming with opinions, if that sounds helpful. To me, it is not. She seems to think she knows best in all matters; that she alone can change the tides of destiny."

"I see how that could be a bother."

"A bother! It has tested my character, *m'ijo*!" Sarita's breathing grew shallow and quick. Her room was washed in light now, and the air had become a bit too warm. She turned to the window and peered outside. Two stately trees stood slightly apart against the enormity of the full moon. She recalled now how she had found La Diosa within the shadows of one such tree. She shook her head, recognizing the tree and bewildered at the sight of it. Had it always been in her backyard?

"Has it?"

"What?" Sarita jumped, gazing back at her son.

"Has it tested you? Has she tested you?"

"Oh, Lala. Yes. Her self-importance tries me. Her arrogance leaves me breathless!" The old woman gasped in frustration. "I sometimes feel like . . . like *grappling* with her."

"Grappling?"

"Seizing hold of her and shaking the stubborn creature until she truly sees me. It seems she sees nothing." Sarita suddenly had the biblical image of Jacob in her head, wrestling with . . . with. . . . She blinked, and the image was gone. "Even don Leonardo is distracted by her," she said, "and few men are less susceptible to distraction than he."

"Few men," Miguel agreed, nodding.

"I suspect he is taken by her beauty." She shrugged, suppressing a yawn. "Men . . . well, after all."

"She's beautiful then?"

"Attractive," she offered with another shrug. "She has attributes, certainly; and a woman with attributes can be . . . formidable." Sarita took a quick breath to clear her head. "I was such a woman."

"You *are* such a woman."

"Beauty is no excuse, in any case," she continued. "The woman's arrogance makes her single-minded and . . . and menacing." Sarita paused, then squinted, as if to focus. What were they talking about?

"You're afraid of her?"

"Of course not! She is trivial, a minor annoyance." She noticed a glint of fascination in her son's eyes and felt confused again. "She may be . . . unusual; yes, unusual. She has a spellbinding aspect, much as your mother did when . . . when . . ."

"As you still do, *Madre*."

At that, Sarita slipped into quiet reflection. She looked out the window again. The moon had not moved. The two trees seemed larger, however. They were perfectly matched and serene in each other's presence—dignified and grand. The stillness of the night only intensified that grandeur, as immense branches reached heavenward, outward, and toward her. As she regarded them, one of the trees seemed suddenly to fade, like a vision made of mist. Sarita closed her eyes. When she opened them again, both trees were again clearly etched against the brilliance of the moon.

"You say she is bothersome," her son was saying, "a minor annoyance."

"No more than that," she concurred.

"To the dream of humanity, she is much more than that."

"To the dream—?" Sarita hesitated, sensing danger. Although still groggy, a part of her became strangely alert to what was happening. She sensed that Miguel, sitting quietly beside her in comfortable conversation, was about to tell her something she was not willing to hear. Her body was sleeping, she realized, but this particular dream was now demanding her full attention. Her son leaned in and spoke to her.

"She's both demon and angel, enemy and ally to human awareness," he said.

"Demon?" she repeated, and was surprised to feel that the word caused her shame.

"And angel."

Sarita stared at Miguel, whose expression was curiously impassive.

"We submit to Lala at our first words," he continued. "She possesses us, mind and body. What is a demon, really, but someone with the power to possess another? We are elevated by her glory and we fall at her whim. With awareness, we can finally see her, confront her—wrestle with her if necessary—and negotiate our own freedom."

"*M'ijo*," she said with foreboding, "you are saying . . . I am *possessed*?"

"Having eaten its seed, every human has the tree of knowledge growing in his or her head. *What sort of fruit does it bear?* anyone may ask himself. Does it cause fear, or respect? Does it taste of nectar, or of bitter poison?"

His voice echoed in the silence. Sarita wanted to wake up, but her body refused to respond. "Lala . . . possessor?" The old woman worried over the words. "She appears to me as nothing of consequence. She is a pest; that is all."

"In the desert," her son said tenderly, "the devil was frightening—formidable, you could say."

Sarita's fingers fidgeted with her nightgown.

"The devil in the garden was not so scary," Miguel continued. "He was mostly annoying . . . and easily crushed."

"That is not a story to ridicule, *m'ijo*," she admonished.

"It's the story of every single human. Perception changes as awareness grows, Sarita. How you perceive Lala now is the measure of your awareness."

Was this woman the devil, then, or was she not? Sarita suddenly felt too old and too tired to sort it out. Why could she not wake up?

"The devil is a distorted image of God," her son said, answering her thoughts, "just as knowledge is a distortion of truth."

"I would not knowingly keep company with evil, my child," Sarita said softly.

"Evil is the result of believing in lies, and it has no chance with you. When it comes to Lala, you see her very much as she is—something beguiling, beautiful, but not deserving of your faith." He took her hand. "You see her only as a nuisance."

"But I sometimes see her as—" She stopped, embarrassed.

"As Sarita?" he prompted, but she said nothing. "Sarita is not the truth of you. Just as this Miguel, as you perceive him, is not real—although, admittedly, he is adorable." He kissed her cheek in the dazzling moonlight.

"Real or not, I love him," she whispered tearfully.

"Alive or not, he will love you forever."

Something scraped against the windowpane, and Sarita came fully awake. Her first impulse was to look for the two trees, but nothing loomed against the night sky but the roofline of her neighbor's house. A young jacaranda was blowing in the wind, brushing its budding branches along the eaves. Turning back again to survey the darkened room, she sensed no other presence.

Alone in her little bed, Sarita breathed deeply and collected her thoughts. She had traveled great distances since the morning of her son's heart attack. She couldn't say where she'd gone or how she'd become so enthralled, but all of it had somehow brought her back to this same spot. She was here at home, and her son was not with her. She was old and must soon face that fact. Gone was the esteemed *curandera* who healed the dying. Gone was the daunting enchantress who shaped fate to her will. Even in trances, she could not bring that woman back. She could still take an inward journey, however. She could still wield the forces of desire. She could still dream. . . .

Sarita turned her gaze toward the window again, contemplating the windswept night. She began to understand how she had come to find her son in the Tree of Life. She did not need Miguel to explain that love, unconditional and unremitting, is the essence of life. She knew well that love based on conditions is a twisted copy of the truth, and yet it is there that humanity plays—there, in the shade of a metaphorical tree that wavers like a sullen mirage and scatters the seeds of a million lies.

No, he did not need to tell her these things . . . and yet her own selfish demands had defined this mission. Sarita turned on a bedside lamp. She rose from the bed, found her slippers, and shuffled slowly to the kitchen, hungry for the taste of *pan dulce* and a cup of herbal tea. She must find another way to dream this. She was seeing her son as misguided, even stupid, concerning his destiny. And yet . . . was he not happy? Was he not the very picture of happiness, now that he had freed himself from the image of Miguel? Sarita wondered if she herself had time left to taste such freedom.

She wondered some more as she nibbled on a piece of sweet bread and warmed her fingers around her hot cup of tea. Placing the cup on a small table in the living room, she settled into the big chair. Still wondering, she drifted into a deeper sleep, allowing this moment—all moments—to dissolve into absolute power.

ARE YOU HAPPY OR ARE YOU STUPID?"

"What do you mean?" the woman asked, her eyes glazed with tears. The bus they were in was bouncing along a mountain road in the Peruvian countryside, engine roaring and gears grinding. Around them, everyone was talking loudly, and she was unsure she'd heard him correctly.

"Are you happy?" don Miguel asked the student again, his eyes twinkling under the brim of his well-worn hat.

"Not . . . not really," she stammered.

"Then . . . you're stupid?"

"No."

"No?"

She tried to look directly into his eyes, but it was difficult. It was hard to meet that open gaze, and to remember exactly what unhappiness felt like.

"Happiness isn't a choice," she finally said, concentrating on the way her hands curled together in her lap. "There are so many things that make us unhappy," she added, sniffing. "There—well, there are people who break our hearts."

"Is that what you believe?" he asked with overstated sincerity. "That things make you unhappy? That someone can break your heart?"

"Yes."

"Someone can make you, force you, to be unhappy? Really?"

"Yes. Does that sound stupid?"

"It sounds like you believe it."

"I can't help it." She shrugged. "Life sucks."

"Life is full of choices," Miguel said sweetly. "Do you want to choose unhappiness?"

"If we really do have choices—then of course it would be stupid to choose unhappiness."

"Exactly."

"But . . ." In the effort to find an argument, it seemed she was again overwhelmed by sadness. She shifted in her seat and began to sob.

"So, honey," Miguel repeated gently, "are you happy, or are you stupid?"

This time, the tears fell abundantly into her waiting hands.

From the seat across the aisle, Lala was watching, listening. The bus was filled with students, all talking and laughing so boisterously that it was hard to hear anything over the noise; but she wanted to be there. She wanted to observe the teacher in action. She wanted to finally understand something about him . . . about the game he played with people's minds—a game he played in much the same way she did, but with a magic that eluded her. She couldn't restrain herself. She wanted to know.

Miguel was eager for this journey, she could tell. The brief marriage that had begun on a summer day in New Mexico had

· 214 ·

been annulled just weeks earlier. Lala could see that his emotional wounds had healed, that he was now eager to teach, to tease, and to flex his power again. So here he was, with a group of zealous followers to join him in the fun. There were forty of them in all—forty exuberant humans on a power journey through Peru, forty people speeding along the serpentine roads that ran at the top of an improbable world.

They were different ages and came from different backgrounds, but they all had one thing in common: the quest. They were looking for solutions to some unnameable problem. They were looking for something they thought they lacked, that had been denied them. She never could get used to the way some humans chased mystery as if it had a higher logic. In the end, they would crave the old familiar knowledge, and they would come back to it. Like roosting chickens after many exasperating attempts to fly, they would come back to routine habits. Sooner rather than later, they would come to their senses.

So why did they bother? Why all the fuss and scratching, all the inspirational singing, the prayer, and the pretense? Was it for this fleeting fellowship . . . or was it for a few moments of exhilarating wonder? Perhaps to come to the edge of knowable things—to gaze, awestruck, into the void—was the thrill they wanted. Once they peeked over the precipice, once they felt the danger, they seemed eager enough to return to the comfort of reality, and to certainty. They went back to what they believed they were, and to the old thoughts that bedeviled them. A little danger, a small risk, was more than enough.

Lala glanced at the adventure-seekers around her. Most of them were Americans, with a few students from Mexico and a couple who had traveled from Europe. Peru was considered by most of the world to be a sacred place, a dream come true for the spiritually adventurous. Mother Sarita certainly believed it to be so. The old woman had sent her son here to bond with

Earth, just as she had insisted he go to Hawaii. She had more places in mind, too—Egypt, Tibet, Antarctica. Each required a different strategy and a road map into a different dream. Perhaps Earth dreamed, as all bodies do; but it also held the memory of human dreams, ancient and forgotten. Aspirations built cities and crushed civilizations, but in the rubble left behind still lay the vibrations of human thought. Knowledge would forever seep into Earth's soil like melting snow . . . and spilled blood.

Someone near her had finished a long joke, and laughter rocked the bus. The excitement on this ride was palpable—even a little nauseating. As the laughter subsided, some women in the back of the bus began to sing, and soon the mood became sickly sentimental. Lala was beginning to regret her decision to tag along. The air was growing so heavy with emotion that it was hard for her to breathe. She was considering her options when something across the aisle caught her eye.

The student was lying across Miguel's lap now, exhausted. Lala shook her head in agitation. That woman had every right to be unhappy. Unhappiness was the natural result of having rigid and principled opinions. After a full day of listening to ideas bang and clash in her head, she was bound to be confused. Misery was inescapable; it was the fate of every human to be besieged by runaway thinking and to suffer for it. Lala looked at the woman again. At the moment she wasn't suffering, apparently. Her unhappiness seemed to have washed away in the deluge of tears, for she now seemed quite at peace. Her eyes were closed, and a smile graced her lips, a smile that was vacant, free of any message or implication. Her expression seemed almost otherworldly—a disagreeable sight to witness.

Miguel began showering the woman with imaginary kisses, delivered by bunched fingertips that pecked at her cheeks, nose, and chin. As Lala watched in fascination, he lifted the woman's wrist and stroked the tender flesh at the inner bend of her arm. Then he flicked his finger over a vein, as a nurse would just before giving an injection, and struck the spot with two fingertips. His nails dug in, and the woman gasped. This was indeed an inoculation, although Lala could not comprehend the science of it. It was as if he were injecting the dazed creature with his own essence: a force that had no status in human understanding. The result of this action was total insensibility. She was quiet, swimming in apparent bliss. Perhaps she was getting a breathtaking glimpse over the edge, but she would soon step back, Lala knew. They always stepped away from mystery; they always returned to certainty, and to her.

Lala had observed a great deal of inexplicable behavior on this trip. She had initially scoffed at the idea of a "power journey"—just one more pretentious phrase in a haze of self-delusions. What did he know of power, this man who subsisted from month to month, lecturing in people's homes and telling tales of truth and awareness? How could he wield power, when he remained anonymous to the world, avoiding attention and controversy? What did he know, this man who laughed at the beliefs that propelled other lives? He could have risen to places of supreme importance in the world, but he did not. He understood power differently, it seemed; but what power could he offer these disciples? If he failed to bring good fortune to the forty who now followed him, would they still see this as a journey of power?

Those were her initial impressions; but the more she watched, the more she wondered . . . and wondering was not among her many talents. She wondered at his strength— both his physical strength and his strength of will. Her own strength seemed to weaken every day, as Miguel called forth power that had no source, in ways that had no explanation. He used words sparingly. He lured the mind onto a path of safety with a few inspiring words, only to abandon it, leaving it to flounder without thought or direction. Words obeyed him. Words stood down or rose to high levels of inspiration, at his will. Rituals rendered his subjects powerless, according to his intent. Knowledge struggled against mystery, at his command. Lala felt weaker, yes. This was not her realm. Here her voice went unheard and her resolve slipped into nothingness. The universe played in this field. Life frolicked here, and even the *nagual* master was susceptible to its whimsy.

She wasn't sure how he did what he did. She wasn't entirely sure what it was that he did. She only knew what she saw. From the first hours in Lima—indeed, from the first minutes on the plane from Los Angeles—he was casting a spell on his students. Women were charmed; men wished to emulate him. The men may well have been envious of the attention Miguel received from women, but they seemed to enjoy the challenge, imagining themselves as gifted wizards. They followed him closely as he climbed ancient steps and mountain paths. They kept by his side, as they had done in the volcano, ready to be his angels and his defenders. Towering above his small frame, they gave an impression of strength and safety. Even the most vulnerable of men seemed to grow strong in his company.

Lala observed the men in this group with delight. They were the prideful protectors of knowledge. They were impressed with ideas. They had been born with a natural inclination to believe, and a leaning toward heroism. Always ready to explore new lands and launch themselves into space, men were not so stouthearted when it came to stretching the membrane of perception. The shaman understood this; he kept an eye on the self-importance of men. He had his chosen few angels to lead sunset ceremonies and guide the others through their rituals. While women cried, his angels gave comfort. When Miguel produced rainstorms, they sheltered him. When thunder rolled and lightning flashed overhead at his beckoning, they stood unflinchingly beside him. Standing tall, these men were metaphors for his unseen power. They were the monoliths that graced his path.

Of course, Lala was accustomed to the loyalty of men, but there was something in these interactions that she could not fathom. Like most warriors, these men were drawn to danger. She understood that. Like most academics, they were drawn to knowledge. Like all children, they were drawn to the good parent that Miguel exemplified. They reached for other things, however, and seemed to find them in his presence. They seemed to find personal courage, where none had been evident. They seemed to find faith, of a kind that had no definition and no name. They found comprehension in the smallest innuendo. They found authenticity—perhaps only for seconds at a time, but it gave them a memory they could take home with them. They were transformed by the inner journey that Miguel encouraged and guided, an unstructured excursion beyond the borders of reason. In the frenzy, they found peace. They seemed to find themselves.

To La Diosa, nothing could sound more sinister.

The women, too, were clearly searching for the truth of themselves, if such a thing existed. Miguel seemed to hold the mirror that showed them what they wanted to see. He was no towering monolith, but his power could be felt. His joy was contagious, and his wisdom was an inspiration. His love had the force of a tidal wave: all things yielded to it. Watching him in action, Lala was ready to admit that he inspired a yearning, one that even knowledge could not. Plagued by that yearning, his students were often diverted from their quest.

"You are mine" were the words every woman wanted to say to Miguel. More than a man, he was a presence, creating a disturbance that transformed. He was changeable, but they couldn't change him. "What about me?" they would cry. *Me* was an invention of the mind, Lala knew. It craved attention, never heeding the consequences. *Me* was every woman's conceit, every man's idea of himself. *Me* was the tempter, breaking hearts and dividing loyalties. *Me* was knowledge, characterized as a person. *Me* was what Mother Sarita could have named Lala, but did not.

Life, to don Miguel, was full of something Lala could not identify or define; something that came before the initial impulse of creation and would last far after the spectacle was over. To him, truth was the one thing worth experiencing, even at the risk of losing *me*. He saw himself in all things and heard life's music in every sound. In his modest human form, he was the exultant mystery that might engulf all those who came close to him. His home was eternity. He lived there, played there, loved there, while those who kept to their small dreams hardly dared to imagine it.

Lala considered what she had seen and heard since her solitary vigil began, here in this rarified world of mountains and secret wisdom. Yes, stirring transformations had occurred. Mystery was everywhere. Music blasted from radios, from the voices of children; and it streamed melodically through the human mind. Music chimed in the air, thrummed deep in the land, and burbled in the currents of rivers and streams. She suspected that music was more than mathematical, more than measured rhythms. Somehow, music conspired with mystery—and, as it was with the shaman, the result was a thing dangerously beyond her control.

⊚ ⊚ ⊚

"Sorry to take you away from your music," Miguel called, "but there's something I want to show you."

It had been a long bus ride, but they were finally at the hotel in Machu Picchu. After a rest and a meal, a small group gathered on the hotel steps, leaving some of their friends in the lounge to dance. When don Miguel called, it was wise to respond. They hadn't traveled four thousand miles to miss the big moments with him, and this might be one of them.

The hotel stood on a cliff directly above the ruins of the ancient city, a power spot that drew spiritual enthusiasts from around the world. After visits to Lima and to Cuzco, the group was spending a few days here, exploring the mysteries of the place and learning about themselves. The lessons began at breakfast each day and continued into the night.

This night in late October, they had all planned to party, but don Miguel was restless. There was more to do, more to say. So, huddling together, the dedicated students quieted down to listen to the master. It was early spring in Peru, in the southern hemisphere, and the nights were still cold. Mountains loomed in every direction, their snowy peaks glistening white in the moonlight. The ruins weren't visible below them, but the night air reverberated with secrets, as if the ruins themselves were recounting tales of a shared dream.

"It's beautiful tonight, isn't it?" Miguel said, pacing the little courtyard.

They nodded. Although most were already shivering, they were silent and alert.

"A pretty moon," he continued. "Wonderful stars! Such a clear and amazing sky!" Saying the obvious was not his usual habit; they suspected something was about to happen. "In a moment, all of this will vanish. Do you want to see how?"

Ah, that was it! A bit of sorcery. A few clapped happily, and Miguel's big angels stood carefully at attention, ready to observe and to learn.

Miguel moved away from them, stepping farther into the

empty night. He began what seemed to be a silent communication of some kind. He spoke no words, and his arms hung by his sides, but one fisted hand moved in slow circles, as if it were twirling an invisible ribbon. Then he returned to the waiting group with a huge smile. No one said anything. The world was silent. Just as silently, a wisp of gray mist crept toward them. From out of the darkness it moved, thin streams gradually billowing into heavy fog. Within seconds everything had been enveloped in an obliterating cloak of vapor. Moon, stars, and mountaintops were gone from sight, and the night was cast in gloom.

"Wow! Amazing, isn't it?" Miguel marveled. His audience agreed, offering him nods of encouragement, although they seemed unable to shape actual words. He took deep breaths of misty air, admiring his own handiwork. "Shall I send it away?" he suggested after a few minutes.

There were no protests. Everyone stood very still, making sounds of consent. Eyes wide, they marveled at the fog around them and waited expectantly.

Miguel stepped away from them again. Without gestures or fanfare, he stood within the mist. He did nothing outwardly noticeable, and said nothing. Little by little, though, the clouds dispersed, becoming thin tendrils of fog that retreated into ravines like peevish ghosts. Everyone laughed when the bright moon burst from obscurity and stars began to wink as if they'd been in on the joke. Within minutes, the mountains shone against the horizon and the world was visible again.

Miguel turned to his students with another big grin. He approached them and began speaking about intent. His voice was almost too soft against the background of live music inside the lounge, but he had everyone's attention. They listened as he explained how intent was life, in conversation with itself. How could such a discourse take place with all the noise and confusion that afflicted the average mind? Surrender was the key to understanding intent. He talked on, as the night sparkled and the dream of ancient warriors hummed in the darkness. His voice was calm, mesmerizing. The force of him was undeniable.

Lala stood by the railing of the veranda, only half-listening. Like her human counterparts, she was struggling to make sense of what she was watching. She had seen a great deal on this journey, and something about it all had rendered her helpless to meddle. While Miguel was like this, mystery outranked her. She had seen this man draw pictures in the clouds, to the astonishment of his followers. Many times she had seen him call for a storm, and the thunder had answered. Lightning split the sky at his command; rain fell at his pleasure. Once, she had joined a few of his men as he led them through hilly terrain to a sacred site. They followed, as he kept ahead of them, and eventually he disappeared from view. Then they followed his footprints. When his footprints also vanished, they collapsed on the ground and dreamed together, having decided that that was the best way to find him. That vision had stayed with her, causing her to marvel at the sort of mind that recognized normal boundaries but continued to reach, and to imagine.

She had seen Miguel appear in two places at once, a phenomenon that amused his more observant students and seemed to require no explanation. "Hey, don Miguel," someone would say, finding him in the middle of an impromptu class on the hotel rooftop. "Wasn't I just talking to you in the lobby?" On occasion, two different women would swear he'd been with them on the same night in two different cities. There seemed no need to draw conclusions. Conclusions were pointless in this universe that they had agreed to share with him.

Lala frowned at the idea of breaking established rules; she was certain that rules were all that held the human dream in place.

Earlier that week, a young woman had fallen while walking down slippery stairs within a cavern. Her back hit a stone step and cracked loudly; immobilized but clearly not devoid of feeling, she screamed in pain. The shaman asked her to look into his eyes, and when she did, she was calmed. Then he took both her hands, lifting her upright, slowly and carefully. Something in her back snapped audibly into place. She was in tears still, but her pain was gone, her mobility restored.

Such tricks and theatrics were familiar to Lala, and yet there was no showmanship involved. Things happened, they were accepted, and then they were forgotten. Anything was possible . . . because anything was possible. That was all. Was this really sorcery, then? What kind of magic resolved fear and stole from the mind its righteous supremacy? She knew the magic induced by fear, by dark thoughts and dire imaginings. She knew well the black magic that moved most of humanity; it was not at work here. Then, what was?

Throughout the human dream, knowledge—secret and selective—had been hiding behind magic. Uneducated masses were traditionally exploited by the few who knew things. Knowledge had seemed magical from the beginnings of human time. It could be that the shaman knew things that others did not. It could also be that he had found a way to make knowledge dispensable. Certainly, Lala felt unnecessary in his world. Since she'd been here, she'd heard no calls for help—not to her, or to God, or to the usual saints, guides, and random entities. All stories were put aside in obedience to life. Her uncertainty stopped her, rattled her, and she found there was nothing to do but watch, notice, and reflect on the nature of something called *white magic* and all the disparities of power.

I feel Lala's gentle intrusion now, as I have throughout my existence. Knowledge follows us everywhere, like a concerned friend or a persuasive lover. It's the discreet noise in our head, whose meaning we think we understand. It asks that our ears ignore what we hear and our eyes deny what we see. It attempts to tell our hearts whom to love and what to hate. At its most intrusive, knowledge is a ruthless autocrat. It will abuse us and demand that we abuse others. One thought can take us far from our normal instincts and compassions. One idea can justify atrocities. It's a simple thing to say that we are knowledge, swept from our own authenticity by words and meanings, but not so simple a thing to grasp, and to change. It's challenging, of course, but faith in ourselves makes it possible, even inevitable.

For many years I've heard Lala's words resonate in the voices of everyone around me. On journeys to exotic places, places far removed from their normal realities, my students take their knowledge with them, as bulky and cumbersome as their backpacks. They let it speak for them, argue and explain things for them. Religious theory and cultural mythology fight the war of ideas through them, until their minds are finally willing to give up. On this planet, every place has been touched by human knowledge. A different kind of traveler will see it, hear it in his own voice, and shift its purpose.

Like every other place, Peru, a land that still echoes with ancient messages and traditions, lies under the spell of Lala. I felt her there, as I did on all my journeys, and it's a pleasure to watch her now as she hears our conversations differently and sees herself in my dreaming eyes. Within her, I can hear words clashing, as they did in Miguel—eventually breaking their own spells and committing themselves to reflect life's great generosity.

Miracle is a word that humanity is drawn to. Miracles are things we cannot explain—at least that's the point of view of knowledge. Miracles, from life's point of view, are occurrences, whether or not we expect them or understand them. First there is nothing, and then there is something; this happens continuously, endlessly. Magic is creative power in action, and it's unfortunate that our human powers include the ability to wreak havoc on ourselves. Black magic is the art of self-defeat. We poison ourselves with judgment and fear, and then we spread the toxin to living beings all around us. To heal ourselves requires self-love, the white magic that works wonders on the dream of humanity.

Knowledge is unsettled by the idea of power. We see how it works in the worlds of business and politics, and we suspect that it works the same in the spiritual world. We presume it's a gift for the exceptional and the few. *She can do it, but we cannot,* people might say to themselves. *He is the chosen one; I am not. He's a master, but I can never be.* We have become masters of what we are *not.* We have made ourselves vulnerable to the belief that others have greater power than we do, because we won't acknowledge the power of us—the truth of us. Power, to the world-dream, is something small and self-serving. Power, from the point of view of creation, is infinite and selfless.

Perceiving ourselves as *life* alters our relationship with everything. Words are usually spoken in anticipation of an action, but they can be spoken *as things occur,* as if the two sides of the equation were in communion. "Do you want to hear the thunder?" I might ask. Boom! They hear it. "Do you want to see the fog roll in?" There it is! "Do you want to see the stars again?" There goes the fog! Wonderful! Which comes first, the question or the response? They are both the same, coming from the same living being. Intent fuels the discourse, and the result is power-in-action. Intent is life; it's the current of life that runs through us, and we respond to that current. Our story is not the truth. Our best scientific equations are not the truth. Symbols can't replace truth, but they can serve it. They can point us toward truth, and when they yield—when knowledge surrenders to what it cannot comprehend—we become the instruments of intent.

Awareness wins the war of ideas. Love wins against self-judgment and personal suffering. The day would come when winning the war would become the central theme of my teaching. That shift had not yet begun during those early pilgrimages to Teotihuacan and power journeys to Peru. Then, the focus was on helping my students get out of hell. It was important for them to know what kind of personal nightmare they had created and how they could finally wake up. They couldn't imagine a way out of hell. The rules had to change, their words had to change, and the inner voices had to be calmed. These students needed to forgive, to be kind to themselves. They needed to see that human existence wasn't just noise and confusion, and that wisdom could be theirs. In me, they sensed truth. In my love for them, they sensed their own power. No one could take them to God. In spiritual terms, I could only help them to find the gates of hell, and to inspire them to move forward, beyond those gates, to heaven.

❁ ❁ ❁

Only three students followed him. The rest of the group had taken a different route, and the bus had gone on to meet them. Earlier, as everyone trekked through the rugged terrain in search of sacred ruins, don Miguel had wandered off the path. The local guide had waved them onward, but there was hesitation among the group. They could see Miguel standing on a distant outcropping of rock, but did he expect them all to follow? With no visible signal from him, they couldn't be sure. After much urging from the guide, they turned up the path, one by one, catching up with the Peruvian and coming together as a group. As they shuffled over the ridge and out of sight, only three stalwarts were left. They kept watching don Miguel, certain that he was summoning them, waiting for them. They began to climb the hill toward his retreating form.

When the three finally found him, the shaman was sitting comfortably on a flat rock. He welcomed them with an expression of interest and amusement. Had he wondered who would follow, who would deviate from expected behavior and leave the safety of the group? Were they the fools, or was everyone else? Second-guessing Miguel Ruiz was a common pastime among his apprentices. It was never enough to follow their instincts and let go of expectations. It was too important to see the deeper reason. *Meaning* was the mind's most cherished prize, Lala knew. If love's tidal wave ever rushed in to obliterate the last, small bastions of reality, *meaning* would be a lifeline for the drowning intellect. Everything must *mean* something. Normally, she approved of the idea, but she recognized its irrelevance here. Miguel was lawless, meaningless.

If Miguel sensed Lala's presence, he said nothing. She sat on the flat rock with the others, her back to the morning sun, and observed. For a while, nothing was said; the three apprentices sat with eyes closed, happy to feel special, pleased to have been chosen. As time passed they began asking Miguel questions, tentatively at first. Lala listened halfheartedly and reflected on the scene. The group was composed of one man and two women. They all appeared to be around the same age: old enough to have growing families, but young enough to withstand the rigors of this kind of work. Assuming one could call it work, she thought. From her perspective, they had signed up for regular doses of insanity. To align with the unknown, to reject the safety of familiar beliefs and traditions, seemed nonsensical. This shaman prided himself on having good sense, yet he confidently led his people to the keen edges of reality, where most of what they had known and trusted abruptly fell away. Some came back to the precipice again and again! How was that not insanity?

"Some people never stop," Miguel said.

Lala turned to him quickly, surprised that he had spoken to her, but he had not. He was smiling at one of the women, who had apparently just posed a question. Lala tried to recall where the conversation had been going. The man in the group had asked something about nature, about the sun and Earth. She couldn't remember his exact words, but Miguel had responded by saying something about "one being." One being, she remembered. Yes.

"There is only one being," he had said, "and countless points of view."

That was it. But if there was only one being, who was *she*? Who was La Diosa, and what was her particular point of view? His voice had put them into a gentle trance, it seemed, and so the conversation faded until the woman spoke. She had been looking at him with wide, wistful eyes, and she said the only thing that mattered.

"What makes you so different from the rest of us?"

There was a pause. The others shook off their drowsiness and turned to stare. The question seemed impertinent.

"Some people never stop," he responded, his smile gone and his eyes clear. He stared at the woman, waiting for another question, but it did not come. She understood, it seemed. Perhaps she was one of those who raced to the edge and dreamed of leaping. If there was a precipice, if it was possible to leap from the bedrock of beliefs that made each person what he thought he was, Lala sensed that this woman would do it. However, there was no such cliff edge. Reality was everywhere, ready to catch those who wandered from their normal thought process and send them back to the point of origin. That, she knew, was the way of human dreaming.

Then the man spoke up, his eyes locked on Miguel.

"Yesterday," he began tentatively, "we were separated into small groups, each one with two group leaders, remember?"

"Yes. How was it?" asked Miguel with interest.

"Well, the two leaders in my group gave us different instructions for the day. One asked us to find the truth in ourselves. The other, trying to lighten things up a little, told us to be entertaining. It was a funny thing to say, I guess, and everybody laughed, but the two things affected me in a huge way."

"Really?"

The man hesitated, expecting judgment. "Yes," he confessed. "In fact, I was so shocked by what I suddenly saw in myself that I spent the rest of the day wandering around alone. Those two ideas exposed a deep conflict in me."

"I don't see a conflict."

"I've met every social challenge in my life by being a clown. Now I can see the conflict between wanting to entertain people—cracking jokes all the time—and being authentic. It seems I've spent my adult life avoiding authenticity."

"You've spent your life creating laughter," Miguel countered.

"Well, if I'm not being a smart-ass, I don't know who to be. That bothers me."

Such concerns seemed nonsensical to Lala, who knew deception to be a simple survival tool. She huffed audibly, but no one appeared to notice. Looking skyward, she saw an eagle riding the spring thermals. She remembered hearing the Peruvian guide tell Miguel that there were no eagles in these mountains. There were condors, of course, that claimed the twisting ravines and open valleys, that soared above the rugged peaks and sparked the imagination of humans. Yet here, cutting through the blue sky and shrieking triumphantly, was a golden eagle.

"I don't see conflict, Chief," he repeated, using the nickname he'd given the student.

Miguel glanced up in response to the eagle's cry. Smiling, he seemed to shift his gaze directly to Lala. At that, her heart stopped, and then the moment was gone. He turned to the others. "You're being true to yourself when you inspire laughter, and you're being true to yourself when you look for authenticity. All the conflicts you talk about hide something simple and true. You are life. At the time of your birth you were truth, in body and action."

"Now I'm a fake," Chief insisted.

"What are you faking?"

"Everything," the man said, shrugging. "Confidence. Awareness. Honesty. If I could fake authenticity, I would." The other two laughed but were quickly silent again.

"With each of you, life made a perfect human being. You're perfect the way you are. You were taught the concept of imperfection, and you built reality around it."

"I was broken yesterday, don Miguel . . . broken."

"And that's perfect. Allow the change to come, and remind yourself that the story you tell about it doesn't matter."

He paused, quietly considering each of them. The women were ambitious, each in her way, but they were cautious about showing it. One was quiet in most situations, but deeply determined. He would use that. The other was more likely on a path to a new husband, but she could benefit from even that sort of determination. It was often possible to turn simple ambitions into lifetime changes. Chief was not as cautious as his companions, and perhaps not so ready to be wise. Miguel had seen great potential in him from the beginning, but he must choose patience as a strategy. Always patience. . . .

From where she sat, Lala could see the back of the man's neck—a brief landscape of sweat and tension. His posture spoke of impatience . . . or was it yearning? Chief was a name inspired by his real-life role as chief executive of some company or other. On this Peruvian journey, Miguel had called upon him to assume the role of a tribal chief, guiding the other students in frequent ceremonies. The name suited him well, but he was a challenge for the shaman. Chief was particularly uncomfortable outside the normal realms of thought. He adored his teacher but often used him as an excuse to be afraid. Once, when challenged to see the truth of himself in the eyes of the shaman, the man had envisioned a grotesque, horned demon with several heads. It had been amusing for Lala to watch, but the experience had sent the warrior to bed with the terrors. He was a determined man, but at times too zealous. At one sunset ceremony he had become so caught up in shamanic fervor that he'd passed out and had to be carried back to the bus. She considered the man and marveled. What kind of warrior swoons? What kind of *nagual* man, as Miguel was fond of calling him, falls victim to his own spell?

At fifty, Chief had achieved success in business and family and was drawn to intangible pursuits and matters of the soul. Competitive by nature, he approached his newfound spirituality with the same tactical fervor as he did any other business. The man was headstrong. His aversion to self-reflection was evident. His opinions were more robust than his spirit. He was an impossible student, really. She liked him a lot.

"I see that you guys are being honest with yourselves," Miguel said. "Your awareness is growing fast. Remember not to believe everything you think—or what others think. Just listen, and learn."

"Isn't there more we could be doing?" one of the women asked. "Chief said he was broken. It seems like we need to keep breaking and breaking to finally shift awareness, and to let some light in." She frowned, apparently unsure if she was on the right track.

The shaman's instinct said she was. "Go ahead," Miguel urged.

"I just wondered if you could—"

"Could you push us harder?" Chief jumped in, as if he had been aching to say this for days. He had said something of the sort before, and apparently forgotten the uncomfortable consequence. "We're here with you, don Miguel, when nobody else followed," Chief blustered. "We're here, and we're ready." The three of them looked at the shaman.

Miguel gave all three of them a hard look. "You want me to push you?" he asked.

"Oh, yeah!" Chief exclaimed. "Do your worst, boss."

"Are you sure?" Miguel asked.

"Have no mercy!" Chief answered with a broad grin. The women seemed less certain now, hardly able to meet Miguel's gaze.

"Do you know what you're asking?" the teacher inquired softly.

Chief kept grinning, but hesitated to speak. Having said the words, he was now being forced to consider their implications. What exactly did it mean to be pushed? In his work, he pushed people on a daily basis. This was different, of course, and he knew the results would be different. Remembering the horrible blue monster of his vision, he stopped to reflect. During the long silence, his smile slowly faded.

Lala, watching with interest, began to comprehend the shaman's point of view. While he looked into the faces of his beloved students, he heard only one voice. He would say it was Lala's voice, no doubt. He would recognize the conceit of wanting to finish the race ahead of everyone else. He would blame her for this constant, urgent demand to know. For her part, she failed to see what might be gained by pushing these apprentices harder. Breaking the human mind was never her strategy. The game she played could be won only by guiding and persuading, then watching the mind do its work. This apprentice, this earnest man who wanted enlightenment to proceed on his schedule, would not break, in any case. He was built to bend. He was inclined to swoon.

The eagle, which had been circling above since Lala had first seen it, suddenly dived toward them, shrieking. The sound ripped the moment and ruptured the sunlit air. All three students felt the shock. To Chief, the man who had demanded to know, it seemed that time was being sucked out through the sky—and that the rugged landscape was rushing in its wake. His mind went vacant. He wobbled where he sat, then slumped to his side, folding naturally into the curves of the rock. The women sat still, breathless, staring at Miguel.

"The mind shifts with small abuses of truth," he said. "A new idea is sometimes enough—a change of perception, or one brief look into itself." His voice was soft, but carried force. "The biggest shifts come from experiencing the pure sensations of life, without commentary. Stop thinking, and there's only sensation. Stop trying, and you rise in love."

Warriors die trying, he wanted to add, but they would likely misunderstand. Spiritual seekers often spend a lifetime trying, bluffing, and strategizing. They fight their inner voices, starve their bodies, and then punish themselves for not finding nirvana. They rummage through boxes of secret knowledge, labor over the puzzle, and still miss its essential piece: surrender.

The wind riffled through the canyon behind them and stirred the last, dry vestiges of winter. It moved the grasses but shattered nothing. Beliefs could be broken without breaking the believer. "It's time to dream," Miguel said, looking into their eyes. "It's time to let go." He paused to make sure they were understanding. "Are you ready?"

He didn't need to ask twice. Each of the women found a place to be comfortable, one on either side of Chief's still-inert body, and allowed her breathing to return to a slow and steady rhythm. Miguel pointed up to the sky and asked them to dream the dream of the eagle that was now soaring above them, drifting ever higher into the heavens. Taking one last glance at the sky, they both closed their eyes and surrendered to the dream.

In the ensuing silence, Lala could feel the currents of something that made her uncomfortable. It rushed into the void that words had left, and vibrated through matter. It pulsed through these dreamers, lying under the sun's enchantment, and it tempered the very fabric of things. Again, the shaman would have a name for it. Perhaps he would call it love, a jewel she worked so hard to tarnish. Love swept theory aside and shamelessly broke the rules. The members of this little group had been practicing the darker arts for a lifetime and were as uneasy about the word as she was. Love was the most dubious kind of magic, dispelling fear and allowing light to pierce through the smoke of knowledge.

What was this ride she was on? What had taken her so far from the shadow of that tree and the grand illusion of words? Where was the old woman, the mother of the *nagual* master, who had pulled her into this mad chase? The eagle cried again in the distance, and Lala was suddenly wary. She looked up, shading her eyes against the sun's merciless blaze. She saw it now—far above her, but somehow recognizable. It seemed that Sarita had found another way to hunt. Lala watched, trancelike, as Sarita soared and circled above the human dream, above the war of ideas, and far beyond the threat of consequence.

Had Sarita soared even beyond her reach? No, there could be no success for this mission without La Diosa. Without her help, there would be no return of the *nagual* teacher, no more words of wisdom from the thirteenth child of the Mexican witch. She'd had enough of dreams and mysteries and deliberate misuses of faith. She hoped the wraiths would never return—Leonardo and his lunatic father, Eziquio. Wherever they were, however they plotted to meddle in her affairs, she was done with them. She was done with them, and with their kind.

What about me? she heard a voice say, and found herself agreeing with it. She must make herself known to this dreamer, and bring him back on her terms. She had concerns of her own, and it was time they were addressed. Lala stood up, took a breath from life's infinite reserve, and calmly let herself fade from the scene, leaving the little humans to their little dreams.

15

S HOW ME HOW YOU GET GOD'S ATTENTION!"
Miguel was calling to the group from the steps of the
Incan ruins. Below him, all forty students froze where
they stood, confusion written on their faces. They were standing at the summit of a small island, tired but exultant, and
behind them, Lake Titicaca glittered with a million opalescent
hues.

Having departed from the mountain shores early that morning, their ferryboat had reached the island just before noon,
and they had promptly begun the hike up to the island's
rocky peak. It had already been a long day, and Miguel was
counting on fatigue to break down the apprentices' resistance
to truth—just a little. Ask someone with a sharp mind about
God, and theories will pour forth. Ask someone who is too
weary to theorize, and you may see the first glimmering of
awareness. *Show me how you get God's attention!* He had given

the command, but nothing happened. No one knew what to do, how to act. Careful not to be seen looking at each other, they waited for the first person to break the silence. One of the men finally did.

"Yo, God!" he bellowed, and the hills echoed his words in pulsing waves.

A few people laughed, and then slowly others responded. They moved in awkward, aimless ways at first, but inspiration soon kicked in. A woman threw herself upon the ground, writhing and moaning with sensual pleasure. Men fell to their knees. Some danced. Many chanted. Chief had apparently had enough and was loath to draw attention to himself; he simply watched the others. One woman placed herself at the center of the group, standing still amid the chaos, and stared up at Miguel, eyes full of purpose. Yes, he thought, that's close, but not close enough. Had any of them thought of a mirror—even a make-believe mirror? Had they felt the urge to stop and be silent? Had they considered directing some attention to the body they occupied? There was no right response to Miguel's question, but it was interesting how quickly they turned it into an intellectual challenge. He could almost hear their thoughts. *Where was God, after all? Who was God? Was he seeing them? Hearing them? Did he care?* How strong was their certainty, their doubt? Did they even like the idea of God? How much did they fear it? Perhaps they feared his judgment, as they feared each other's. Then again, perhaps God's attention was like theirs— weak, fleeting, and susceptible.

Miguel watched in silence for several minutes; then he shifted his attention deliberately to the lake and to the endless skies above him. An eagle cried into the wind as it dove toward the water. He shook his head, smiling. Every day had brought a glimpse of this eagle, doubtless the same one. He sensed intent behind its appearances, but couldn't be sure the bird was even real. The creature seemed like a messenger from another time—a time that had not yet come, perhaps.

The morning had been all about dreaming. Before they'd started the climb to the island's topmost point and to this sacred site, he had divided his students into three groups. Each group had been directed to dream the dream of a power animal. One group would dream like an eagle. This group would feel important, he knew. A second group would dream the dream of jaguars. They, too, would feel special. The third group would dream the dream of a spider. It was easy to see the disappointment in the eyes of those who were chosen to be spiders. *A spider? How little he must think of me!* he could almost hear them thinking. Their feelings would change by the end of the day. Every creature was a warrior in its own right. A good dreamer would see this.

Each of the three was a stalker, a predator. Like humans, they were designed to survive on the life force of others, but in superbly distinct ways. It was obvious that eagles had an advantage over most—soaring to great heights, with powerful wings and a body that seemed almost weightless. It's an ecstatic experience to dream yourself as an eagle, and a terrifying one as well. The air is an ocean of currents and sensory signals to this creature. Shifting frequencies of sunlight create challenges and opportunities. The dream of an eagle offers many revelations that are useful to the spiritual warrior: An enlightened mind knows how to use imagination as wings, flying ever higher toward the unknowable. An open mind allows the human to see from the greatest possible vantage point. An aware mind can recognize its own appetites for inspiration—and for poison.

A jaguar, too, is a killer. Beautiful it may be but, it is stealthy and dangerous. Its prey is life, of course, but what also did the human mind stalk, and why? Was it really so hard, Miguel wondered, for even the gentlest woman among them to imagine herself as a stalker, a hunter with a thirst for blood? Both men and women needed to see what a supremely clever predator the mind is. What the human animal could not always do—bring down the bigger prey, for instance—the human mind did routinely. They would need a chance to reflect on this—all of these hunters. Early in its education every mind learns how to use words to threaten, to punish, and to destroy. If that fact sounded harsh to his apprentices, he had only to remind them of how they spoke to themselves—how they attacked themselves on a daily basis. The mind is its own kind of predator, whether it belongs to the body of a bashful woman or a brawny man.

Miguel glanced at his students, still shouting and dancing, still obeying the mind, which said, "Be right, be clever, be the best!" Given enough time, they would give up. Looking for God often did that to a person. There are places the mind cannot go, but the body can. Within the smallest veins of the body, within its atomic structure, is the mystery that humans crave. Every heartbeat pounds in response to it; every breath absorbs the truth, and exhales it into the universe. Where else should they look for comfort? Whose attention did they really need? Already, some had hit a mental wall and collapsed where they were. Some were lying curled up in the dirt, murmuring softly. Some had stretched out on their back to stare at the sky and contemplate the unimaginable. The rest danced on, like children in pursuit of a parent's smile.

He thought of the spider. Now there was a dreamer! Yes, spiders hunt and kill. Like humans, they learn to bring down bigger beasts to survive. It's not always about surprise and attack, however. Like humans, they find ways to lure and entrap their prey. They build something beautiful, something that sparkles with dew on a sunlit morning—something that thrums musically at the touch of a breeze and shivers with excitement at the approach of another living being. The spider's hunting skill is its stunning art. Miguel had asked seven women to dream the dream of spiders. All of them were seated now, relaxed and smiling under a cascade of sunlight. Their faces glowed, their hair stirred sensually in the wind, and their minds were quiet at last. So was the way of the spider . . . to create something that sparkles in the dim, dull world of expected things, and then to wait. God would come to them. More than that, the spider dreams itself as the one who encompasses all things, the one who brings all things together with the delicate pull of intent. The spider lives as if it were God.

Hearing the shrill call of a hunter, Miguel looked up to see the eagle, floating over the island like a stalker of dreams—easily, languidly, with eyes focused and talons relaxed. What did this hunter want? What food did this one seek? he wondered. What secret prize did it crave? The answer was obvious, instinctual. All creatures seek life.

What would this one have to kill to be gratified?

Mother Sarita could see it all. She had taught her sons to do this, to put their imagination to the work of dreaming the dreams of other creatures. Miguel liked to say that to dream himself as a human was his one true art—but he had learned much by experimenting, she knew. To use this kind of power at her age, she had to push away the doubts. She had to push away the noise. She had to forget the body, her family, and the laws of physical reality. She had detached from her own dream completely and let herself fall, let herself rise, let herself fly freely on the wings of imagination.

Now, soaring within an eagle dream, she saw what dear

Sarita, the old woman and rattled mother, could not allow herself to perceive. From this sweeping perspective, she could breathe the long, rhythmic breath of body Earth as it floated in infinite space. She could see Peru, dressed in warm colors, marching boldly along the Pacific coastline, so far away and yet so accessible to her now. She could admire the country's forested valleys and sun-struck mountains, its shadows and light flows, its daydreams and its deep and tranquil sleep. From this vantage point, high in Earth's atmosphere, she could see no traces of humanity. This land, so like the land of her birth, was a living painting, a portrait of something that would remain forever unseen but for its reflection in matter. She felt sublimely grateful to life—for color, texture, and the senses she'd been given to savor it all. She could feel the uplifting arms of the wind, and the way each plume on her body responded to the subtle messages delivered in the current around her. Her big wings pushed against the force of life itself, then submitted and relaxed. Her eyes took in the broad scope of things below, things that shone along the horizon, things that moved and things that did not move yet were there.

She had seen her son leading a group of followers to the sacred sites among the ridges and ravines of the Andes. She had swept so low one brilliant afternoon that she'd imagined she could hear his heart beating as he looked up to her with respect and recognition. He kept his people enthralled—by the beauty of Earth herself, of course, but also by the dizzying love that emanated from him. His love could be felt, even in the air, over distances, and through the crystalline haze that was her dream. His message was love's message. What she had found so irresistible about him as a child caused others to yield to him as a man. He delivered an uncompromising message of truth and awareness, but it was the messenger who made the difference. Words served him, but his power was in his very being. She could see it from the perspective of the eagle, a hunter of life. She could sense it pounding through this hunter's blood and this strong, eternal heart.

She realized so much now, from this expansive viewpoint. Racing over this vast terrain, she could see the strange desperation of humans searching for a truth they imagined existed elsewhere, but which existed only in them. She could see their determination to touch the sky, as if the answer to all mysteries hovered there, waiting. She saw messages etched in the dirt, carved in rock, and wrapped within temple walls. She saw signs of undiscovered pyramids, once proof to humanity's need to burst from the bonds of Earth and reach the heavens. She understood people's desire for union, for connection. They wanted the kind of vision and supremacy that eagles enjoyed. She heard the cries of countless humans, begging for release, howling to be delivered from captivity. She saw the pageant, the parade, as it wound around Earth and spun through human history—people seeking God in everything but themselves. If they could see themselves as they were, it would play out so differently. If they could just recognize themselves as the real and present embodiment of life's mystery. . . .

And what of Sarita, she wondered. Was she part of this eternal dream, this creative force forever in pursuit of itself—or was she simply a human, a woman, a mother in search of a lost child? Did she really need to preserve the life of one man to prove herself? Was her son's life not promised to eternity, either way? Sarita blinked against the last bold rays of a disappearing sun. Had she truly been selfish? Many had called her that over the years, but she had taken care of two husbands and thirteen children, after all. She had seen her babies grow to be strong and capable adults in a disobliging world. She had stopped at nothing to support and protect her own. So whom was she protecting now? She had imagined it was Miguel, the child that had come to her in a flash of light. He was the thirteenth of thirteen, the pride of her ancestors—and hadn't those ancestors responded to her call for help? Wasn't that proof of her generosity?

Perhaps . . . perhaps she was trying to bend this dream to her desires. Perhaps the ancestors respond to Sarita, she thought, and because of Sarita. Perhaps they speak with my voice, and for my sake only. Perhaps salvation means something different to my son—my sweet, wise child, who lingers now on the precipice of infinity, waiting for my permission to surrender to it.

She felt her wings growing tired, and she tipped them toward the eastern shore, toward a well-deserved sleep among the pines. She could not imagine where she would go from there, or what she would do. How could she abandon this task? How could she give up now? What would be the consequence to her son, the loving messenger to an unloving world?

She swooped lower, toward shadows that were gaining courage in the wake of the sun. She would rest, yes. She had witnessed enough through the eyes of the eagle. She had done more than could be expected of a human. It was time to know life in its totality. It was time to see how the view looked from life's infinite perspective, and to abandon the small perceptions of matter. This, as her son had so often said, was the art of mastering death . . . even in life.

Don Leonardo watched the eagle circling ever lower in the evening light. The grace of it—that slow and deliberate spin toward earthly things as it sought a safe and tranquil refuge—made him think of home. Home used to mean a country, a house, a family. Now home was a place that existed beyond human imagination. Now it meant truth and timelessness.

He and don Eziquio were enjoying the sunset skies from where they stood on a ledge among the remains of a temple that had been built by the Incas in a century long forgotten. They both felt the potency of that faraway dream and could almost hear the resonant hum of other conversations, words spoken in ancient tongues and echoing beneath the ruined archways of the temple. This present moment, this sunset, was also part of a dream, one that was foreign to them and yet hauntingly familiar. A *nagual* man was speaking to his loyal apprentices in the courtyard. No one listening now could imagine this place as it had been, or could envision how many such students had sat, quietly listening, in other times. This was where the great masters had come to share their wisdom.

In the present dream, the master was his own grandson, the heir to a legacy that was unlikely ever to end. From the beginnings of time until humans drew their final breath, the search for truth would compel them, and the wisdom-sharing would continue. Even beyond the dream of humanity, there would still be life, striving to reflect itself in all things. The sharing would go on. Don Leonardo knew, as well as he knew the dream of these ancient people, that his kind of sharing went beyond words. Miguel shared himself completely—from body, to mind, to the force of his presence. What Sarita was hoping to do was a disservice to her son. Such a life force could not be taken apart, its pieces tallied and sewn together into something whole, as she had imagined. Such a presence could no longer be contained. Miguel had realized that, even before his heart failed. He had awakened that final night from a warrior's dream, ready to leave the flesh. Life was eager to release itself from the constraints of matter. Was this really so hard for a mother to see?

His eyes darted upward, searching for the eagle, but it had drifted earthward and into another dream. Miguel's students had picked up their belongings and gone, eager now to return to the mainland and looking forward to a warm bed. They seemed exhausted by the challenges of truth. Leonardo chuckled to himself, wondering how exhausting it really was to let a wiser man give you the food you were unwilling to forage for yourself. He shrugged then, knowing how acutely their appetites would need to change even to accept the food that was being offered. In point of fact, they deserved much credit.

Near him, he could hear his father grumbling. Leonardo glanced at Eziquio, who seemed to be dealing with his own challenges. The two of them had enjoyed little opportunity to engage in conversation. To one who was barely more than a reverie himself, it seemed that don Eziquio had been long absent from Miguel's dream. Was the boy reluctant to put the old man's talents to use? Perhaps he was unsure whether the legendary trickster would help—or hurt. Leonardo watched as don Eziquio, having jumped down from the ledge, paced among the ruins and kicked the rocks with his worn boots, looking as if he were searching for something lost.

"It is all too confusing," his father grumbled, rubbing his brow in agitation. "To see a man's life in this way, inspecting one subtle moment at a time, is like counting the seeds of rice tossed at a wedding."

"It is revealing, though, is it not?" don Leonardo suggested.

"Vexing. Like picking pebbles from a pot of black beans."

"More like painting, I think."

"Like finding fleas on a street dog."

"No, like painting," his son insisted, wondering if it was Eziquio himself who was unwilling to lend his talents to this cause. "I was good at it once. As a young man, I often attempted to paint my sweetheart . . . whose lovely name I have now forgotten. I tried many times. Mesmerized with every detail, I failed to see what I was painting."

"You failed to see your sweetheart?" Eziquio stopped and squinted at Leonardo in the waning light. He seemed to be seeing his son for the first time. The musician had aspirations as a painter. The soldier had been a tender lover. What else had he not noticed about the boy?

"I would be dotting and dabbing," Leonardo continued, "mixing colors in the fashion of great impressionists, enchanted by my own genius, and then I would step back to discover that the girl I'd painted was missing a nose, an ear, or a left eye. In one instance, her face was almost obscured by an elaborate orchid. In short, I was infatuated by the medium, not the woman. Do you see?"

"See what? That we are attempting to find a face in a flower bed?"

"We might want to remember our purpose here."

"Which is . . . ?" Eziquio asked, his curiosity piqued. Perhaps the boy had found the lost gem in the rubble. Kicking one more rock, he sprinted up some steps and landed next to his son.

"To see what we came to see," Leonardo answered.

"Not to be infatuated?"

"You are the soul of infatuation, *Papá*." He squeezed his father's shoulder affectionately and was astonished at the feel of it. Even to him, the man was barely more than a vision. "We are here to *see*; that is all."

"The process seems confusing," Eziquio said, "like sorting dreams from delusions, or specks from dust clouds."

"Tricky work, I admit," his son agreed.

"Like separating granules of sugar from grains of salt."

"If Sarita must do it, we must help. If doña Lala follows, not even realizing she is a hunter in love with her prey, then we must follow also."

"Shall I not fall in love as well?" Eziquio asked earnestly, his bony hands pressed to his heart. "I admit that would be a trick, in my condition."

"You are love itself," Leonardo answered, resisting the impulse to touch his father again. "Ready and eager to inspire."

Don Eziquio nodded, wondering for an instant why he and his son had never had conversations like this in their living years. Father and son: that phrase carried meaning of a kind that was invisible and unspoken. His son had not been born his equal. Little Léo had been a ridiculous thing for so long that, when the child grew into a man, his father continued to see him as slight and peripheral. When the man grew into a master, Eziquio was too distracted to notice—distracted by life, family, the challenges of survival, and the desire to rise above this human dream. In the end, he was too distracted by his own games and illusions to see what he had created. Well, father and son were having a conversation now, and there was little left to distract them. He would make the most of this moment, be it real or fantastical. He would unleash his greatest asset—his imagination—and go to work. He would recall a purpose, and if that failed, he would invent his own.

Little Leonardo was telling him that he was love itself. Of course he was! Did he need his son, however wise he had become, to remind him that it was a shaman's pleasure to inspire? Of course not! Eziquio tapped a booted foot and snapped his fingers. He took a few quick steps in each of the sacred directions and then stood still. He closed his eyes and began dreaming the dream of humanity. All he needed was there—within human thoughts, and in the power that made all thoughts possible. Thoughts, plans, purposes—not one of these matters to Earth, the slow-rolling dreamer who nurtures us all. And none of it mattered to him. Between one thought and the next, however, lay a universe of possibility. Between one thought and another lay unending mystery. Yes, don Eziquio was ready and eager to inspire.

Inspiration was often mistaken for the sudden loss of reason, he reminded himself; but *insanity* was his preferred medium. He must call forth all the colors of divine madness and splash them upon a ready canvas. He must find an apprentice to play with. Like a spider, he must spin translucent memories, the kind that sparkled and shone only in the most pristine light. He must strum at the delicate threads that connected Miguel to someone he cared for, taught, and tormented. Through his apprentices, the heart and soul of a master could be found. So, the artist called Eziquio would seek out a student, a particular one, and attempt to stir the life force of an almost lifeless man.

That decided, the old trickster promptly vanished from the scene, leaving his son Leonardo to find another precious piece of the dream and to capture it through his own artistry.

16

WHEN I WAS IN HIGH SCHOOL, AT FIFTEEN OR sixteen years old, I happened to make friends with a girl in the lower grades. She was thirteen, exceptionally pretty, and to my eyes the embodiment of innocence. I felt protective of her and made a point to watch out for her when she was around the older kids in school. It didn't occur to me to compromise her innocence in any way, but it seemed that she misinterpreted my attentions. One day, while we were laughing over something during lunch, she leaned over to kiss me. I was surprised and touched. It was an endearing gesture, but from her naive point of view, it made us sweethearts.

"Now you're mine," she stated confidently, as if it were inevitable, and kissed me again.

I didn't know how to tell her that my feelings were not like hers, that she was a child to me. I didn't know how to do it without hurting her, so I said nothing. As I see it now, telling the truth would have been kinder than my kindest deception.

I played the role of her boyfriend. My friends laughed behind my back—I could hear the whistles and snorts of derision as she and I walked from school hand in hand—but I refused to be the boyfriend who rejected her. The time I spent pretending to be a suitor to a little girl who had put me at the center of her romantic fantasies was a learning experience for me. Having already mastered the usual romantic strategies of a teenage boy—a balancing act of control and casual neglect—I had now put myself in the position of showing respect. It was surprisingly easy, I discovered. I was already jaded at sixteen, but she reminded me of the ideal love that I'd once envisioned as a child. That ideal had been all but lost in my hurry to grow up and have sex.

Our little romance didn't last very long. That summer, her family moved away and she and I parted sadly. The impression she made on me lasted most of a lifetime, however. The thirteen-year-old girl was gone, never to return, but I have seen her in many girls since—girls playing the role of women and wives. They each had an image of me that suited their childish ideals. They failed to see the truth of me, or the truth of themselves.

Everyone is dreaming. Each person's dream is based on his or her own perception of reality. Within this individual dream, there exist many characters—family members, friends, lovers—and they have specific qualities, according to each of us. They make emotional impressions on us, and we react to each of them differently. But we react according to the way *we wish to perceive them,* not according to who they are. How can we know who they are, except as fictionalized characters in our own movie? The life, the movie, the dream—all of it is produced and realized by the dreamer. Walk into a dear friend's movie, and you will see many of the characters that also appear in your movie, but they will seem different, and will evoke a different emotional reaction. The dialogue will be different, and key dramatic scenes won't play out the way you'd expect. Walk into your father's movie, your sister's, your spouse's, and you'll find in each a different kind of story. The character that you play in their dreams may not be as sympathetic as you'd like, or as unflattering as you'd suspect. Your character may not be central to the plot; alternatively, it may have exaggerated importance. In other words, you might not recognize yourself in the dreams of those around you.

Everyone is dreaming his or her life in a unique way. We assume that the dreams of others are just like ours, but they are not. The important thing is to respect the way others dream, even if we don't approve of their interpretation of reality. We can't force someone else to see life as we see it. We wouldn't want all interpretations to be the same, and we shouldn't wish for it. We have a chance to turn our head on the pillow each morning and see the person we love as if for the first time. Just as we would like to be perceived without assumptions or past prejudices, we can allow their unique qualities to impress us all over again . . . expecting nothing, and accepting them as they are.

Everyone I meet sees me differently, according to the way they dream themselves. Who I am has very little to do with other people's opinions about me. I may be heroic in their eyes, or I may seem threatening. The truth of me, like the truth of them, is something impossible to define. As a boy, watching the spectacle of my brother's funeral and listening to people play their expected roles, I realized that all the talk and all the posturing concealed the exquisite truth of them. The image we have of someone else is usually created at a glance, and rarely altered. We do the same with ourselves, settling on an image and working hard to live up to it. We dress up every day as if we were going to a masquerade, clothed in our bravado and our beliefs. Seeing our own play-acting, we can change. We can begin to sense what's going on under the words, free ourselves of automatic responses, and react authentically.

The thirteen-year-old girl that I remember still touches me. I remember her spontaneity and her joy. I remember the trusting way she looked at me as I spoke, and how she hung on every word I said. She was guileless and unselfish, with a musical laugh and a heart-melting smile. She hadn't yet learned to use love against herself, and I can only wish she never did. Her sweetness was a lesson to me—it's always there, in every woman, whether they can sense it in themselves or not.

In fact, I heard the same little laugh decades later, not long after Dhara and I had separated and my brief marriage had been annulled. I was with a large group of students in Peru, and on our first night in Machu Picchu I noticed a woman dancing in the lounge among the others. She was dancing alone, feeling the music and smiling blissfully to herself. She was a good dreamer even then, creating the world she wanted in her imagination, unaffected by the commotion around her. As she swayed to the music, something must have amused her, and I heard the endearing laugh of a thirteen-year-old in love. I saw her as she was, for just an instant, and comprehended the wisdom that is sometimes mistaken for childish innocence. In that instant, I knew that the world this woman imagined would soon come to life for her, with practice and with the power of her intent.

These have always been the interesting students to me—the ones who are in touch with the dreaming mind and willing to step beyond its boundaries. These are the strongest dreamers, the shy ones who demand nothing and are capable of imagining everything.

The journeys to Peru and other power places have been key to my own evolution, not just helpful to my apprentices. Bus rides and boat trips, rituals and rites of passage—all these have added up to moments of increased awareness for me as well.

On one particular journey I fell in love with the woman with a young girl's laugh, and felt as strong an ardor for another person as I've ever felt. Like so many others, she was feeling lost and was determined to find her way back to clarity. She never imagined that the journey would take her far beyond the comfortable illusions of clarity, toward true seeing. She never imagined that she would become a master herself. She never imagined that she would someday have to watch me die, as I'm dying now.

As I observed Emma that first evening, I could see how the student might someday become an inspiring messenger. I wanted to encourage that prophecy, give it the protection it deserved, and see if at least one of these forty could come awake. They were all adults now. They were grown up, physically and intellectually. Now was the time for them to leap past ordinary thinking, to go as far as they could beyond old distractions, and to truly *see*.

With that in mind, and with Emma in my sights, I was energized. I wanted to find new ways to push reasonable minds past the barriers of reason.

◉ ◉ ◉

Beyond reason's edge? Did you think that was possible—or even desirable?"

Lala was sitting in the tree with him, her diaphanous skirts floating around her as she straddled a wide limb and met him face to face, eye to eye. She had willed herself into the heart of his dream, into the Tree of Life. Doing so had taken an immense act of power that first required the distasteful task of disbelieving—disbelieving the dream she was in, yes, but most abhorrent, disbelieving herself. There had never been any question that she was real, but to confront the shaman in this manner, she would have to show herself from a different perspective. She must present herself as myth, a character in his dream. She must make herself deniable, challenge him, and win him back again.

"Reason is my realm," she insisted, "and it is also life's realm. It is built into the brains of humans."

"Nothing is outside of life's realm," Miguel answered her calmly. He seemed not a bit surprised to find her sitting there. He even seemed relieved, and this troubled her. As they stared at each other, the planet Earth moved in lazy circles behind them. There was too much light in this place—far too much, she felt. This light rinsed away distortion like a purge, and nothing reflected it back but an illusory planet, this preposterous tree, and their dreamlike faces.

"Would you prefer to have this conversation somewhere else?" Miguel asked. His gaze turned to the shadows beyond, where she could see the outline of a tree much like this one, but in a landscape that was easier on the eyes. If she was to win at this, she must meet him on his own ground. She must allow him to believe the advantages were all his.

"I prefer it here," she said crisply. "I like your . . . vision."

"Me, too," he said, looking away toward the infinite.

For a moment she thought she was losing him. His attention drifted away from her, to something invisible to her. After a long moment, he was back again, apparently recognizing her face and searching her eyes as if they, too, provided a view of the infinite. She had forgotten how much desire those eyes stirred in her. They sparked a strange craving: to dissolve and to abandon herself. Shivering, she concentrated on the sound of his voice.

"Have you enjoyed my memories?" he asked, smiling. "Have the events of my life clarified anything? Have they offered any illumination?"

Illuminate. Clarify. Was light going to be his weapon, then? Light was simple enough to bend and deflect, she thought. Any substance—even the suggestion of substance—could send it careening into other orbits. Knowledge was her shield, and it would decide the outcome. Lala blinked, refocused, and smiled back at him.

"This is illusion," she said. "You and I know it, and we both appreciate the value of such illusions."

"Which is . . . ?"

"To soothe the mind, to inspire the human. Don Miguel, the man, the son, the shaman, is ailing."

"He is dying."

"His body has betrayed him, and the mind must make adjustments to that fact. This dream is a retreat, a way to rest—logical, but we must take care not to corrupt logic while we attend to the needs of the flesh."

"We?" Miguel was still smiling kindly at her, but his kindness now seemed touched by irony. "There is no *we,* my love," he whispered. "Body or no body, *I am.*" He paused. "I am . . . that is all."

"Do not misunderstand my words—"

"There is no *we.* If my human survives this, he'll have no patience for the game."

If my human survives? So it was still possible. Lala tried harder to concentrate. She was in the shaman's dream and must play by its rules.

"We must . . . we must illuminate things that have been ignored for too long," she said reasonably. "We must clarify, as you say, and make agreements."

Miguel said nothing.

"Look at what you have conjured, wizard," she said, gesturing around her. "There is this terrain, this paradise of simplicity. There is you, right here. There is Earth, the mother of existence, dancing in the dark like a lover."

"There is you," he noted.

"There are two trees in your paradise, and—"

"There is you."

She stopped and looked at him, waiting for a change in expression. What was this emotionless dream that resisted her taunting? Miguel's face showed nothing. His eyes drew her into their depths, and she felt the impulse to leap, to melt, and to forget herself there. Instead, she braced herself for a fight.

"Two trees," she repeated. "Two trees. Do they both not bear fruit? Are they both not fertile? We are both creator and the product of our creation. My world and yours are the same."

She waited. He said nothing.

"Out of a single cell," she continued, leaning toward him, against every well-crafted impulse of her own, "you create a universe of cells, a living being. Out of a single utterance I create a universe of ideas, a living dream. Without that dream of words and meanings, each human universe would be nothing, do nothing, realize nothing."

Still, he was silent. The man could be infuriating.

"Look upon these two trees, exactly the same in beauty and holiness! You think of me as a menace? No, I am a blessing. It is humankind, raised from a bed of algae, that is the menace. Humanity was destined to breathe, to eat, and to become extinct—a parasite on the celestial body to which it clings!"

The shaman watched her with interest, but he said nothing. As her indignation grew, she felt the mood of the place change. There was a feeling emanating from him. No, not a feeling, a . . . power. He had spent his life talking about love as all people speak about love, in pathetic and giddy ways. It wasn't until now that she felt the force of it, the truth of it. That was his weapon, the indefensible argument. People had come to him many times like this, ready to fight and eager to win. They had all succumbed to the force of love, their arguments defeated. Suddenly wary, she was reminded of other times, other confrontations. She had done this before, she recalled, but who had won? Had it been love, always?

"More memories are waiting, *m'ija*," the shaman said. "Resume your search. Recall how knowledge may serve awareness."

That was too much! Rage swelled in her, throwing shadows and signaling the storm before the lightning. Well, what of it? She cared not if the tempest came, if her eyes glowed red and the air turned hot and wild around her. She cared not if this tranquil dream was obliterated and thrown to the farthest points of eternity. She would say again what must be said.

"Look upon me, messenger!" she shouted, her hair turning crimson as it flew with the rising winds. Her body levitated toward the starless skies, the elegant dress spinning bits of singed silk into the darkness like fireflies. "I have lifted humanity from the dung pile! I have enhanced and inspired man since his first choking breath!"

Every leaf on the great tree fluttered, sending gold and silver sparks over her angry face and out to the silent landscape beyond. Lala's fury fed more fury, and the tree responded to each angry word. "Do you imagine yourself to be the essence of life, the brilliant creator?" she demanded. "What sort of genius leaves his creation to toil, suffer, and submit to hideous death? I have given humans the means to survive death. Words survive, and memories persist. Long past existence, traces of knowledge will remain, etched in rock and embedded in starlight!"

"Yes, my love," Miguel finally said.

She looked at him, and her anger turned to horror. His angelic face had transformed into the face of a demon, eyes like burning coals and hair on fire, spiraling chaotically upward and into blackness. The vision arrested her, taking her breath. She saw *herself* in that instant, and gasped.

"We are the same," said her reflection. "We are the echoes of life, stories shouted into deep ravines. Hear how we shout," he said, and the stillness was flooded with a horrible noise. It seemed that billions of humans were yelling, fighting, pleading, all at once.

"Don't tire me with your tricks!" she bellowed, and her words came back at her in a firestorm. The face in front of her parodied her own. The voice that answered her was so alarming that it shook her resolve.

"See me," mocked her mirror image. "Hear me. Yield to me."

"Stop!"

"Fear me. Obey me. Say you are mine!"

Lala, the goddess and the tyrant, could not go on. Her words echoed in her ears, their sound losing all meaning. Her anger was swept away by the wind, leaving her to float downward again, into stillness. She was seated on the limb of the tree again—life's enduring tree. Beside her sat Miguel, the man who spoke softly and smiled.

"It's time to let go," he said.

Was he speaking of himself? What did he mean? "Would you cast your precious life aside," she whimpered, "so easily, so heartlessly?"

Miguel looked at her. "The love of my life is my body," he said. "Should I condemn it to a lifetime of pain—so heartlessly?"

So that was it, she thought, and sighed. She was tired and would soon lose the will to continue. Defiance was difficult now, and thinking had become tedious. As she sat there, discouraged and withdrawn, Miguel edged closer. She could feel him reaching for her. With arms that were implausibly strong and unnaturally warm, he brought her gently into his embrace.

Lala cringed. How could she hope to survive this trespass? How could she—untouchable and untraceable—yield to the caress of life? What was this terrible magic? She fought against him . . . or so she imagined. The more she envisioned herself struggling, the weaker she became. Ultimately, time lost significance and struggle lost its virtue. At the precise instant her purpose was forgotten, Miguel eased his hold on her.

"Go now," he urged. "See what needs to be seen."

"What is left to see?" she whined. "I have observed your life; must I witness your death?"

What voice was this, she wondered, hardly knowing why she'd asked the question. Looking up, she saw lights flickering across the global landscape of Earth, and she sensed how knowledge might still play a part in his transformation. Wasn't that her purpose, after all?

"There are still impassioned words to be spoken, to be written!" she said, remembering the finer aspects of herself.

"And there are others rising up to speak them, even as I lie dying. Go. See."

She sat up, adjusting her gossamer dress with dignity. Her eyes were brighter now, glowing with a subtle fire as she directed one last, wistful look at the tree across the valley, which brooded majestically under sullen skies.

In the next breath, Miguel was alone again, pestered no more by the temptations of knowledge. He took in an exhilarating breath of infinity. As he exhaled, he sensed another shift, one that took him farther away from the curious pull of matter. Soon, he thought. Soon, the end will come.

His attention moved to the sky, the mystery, and the peace that comes with absolute freedom. He had occupied matter and felt its discomforts. He had participated intimately in the human dream. He had watched the spectacle—he may even have caused some of it. It had been intoxicating, but he did not wish to return. With him or without him, humans would go on living and reliving their painful dramas until they chose to wake up.

Awakenings happen to the living. A man can wake up, see the prison he has made, and decide to choose freedom. He can end his appetite for lies and discover the sweet taste of truth. He can persevere, changing his point of view from mind to matter to the dream of light, and then to the exquisite realization of himself as life. The reflection can be redeemed; the mirror image can come within a heartbeat of its maker, see itself, and then shatter into total awareness. It can happen in one man's lifetime.

He knew this, because it had happened to him.

17

D ON EZIQUIO HAD BEEN BUSY. HAVING DIS-
tanced himself from his companions, he could now
exploit his talents. This wasn't to say that he clearly
recalled those talents. He was working on instinct, if one could
call it that. What instincts could a man rely on after having
been so long deceased? The magic he had applied in his lifetime was now a memory, alive for others but not for him. Everything depended on those who had once been close to him,
and how they told his story.

He sensed how his reputation had grown since his death.
He was the sorcerer, an ethereal character in the minds of sur-
viving relatives. Where once his actions had been seen as fool-
hardy and irresponsible, they were now legendary. Yet by all
accounts, including his own, he had occasionally performed
real miracles. He had left even his harshest critics breathless
with wonder. He had created doubt in people who stuck fast

to old ideas like bugs on hot tar. He had moved minds and changed dreams. He'd been tricky, yes, making mischief to offset his own boredom—and, if he'd caused a few nightmares, so what? He'd kept people guessing, marveling, and had shaken them from sleepy predictability. He'd been the master of fun.

Was there anything to a man's life that mattered more than fun? Was there anything between two people, in love or in friendship, that mattered more? Relationships survived on fun. If he and his beloved Cruzita had forgotten to enjoy their lives, had failed to relish every moment together, how might they have lasted so long as man and wife? The same for Celeste. And Esperanza. And his darling Cha-Cha! How would any of his romances have endured so long? Without those many thousand nights of laughter—of enchanting discovery—would he have enjoyed more than a century of existence? Would he have had so much fun? Fun was the point!

Dead or not, he was having fun at this moment. Teachings be damned, he thought, chuckling to himself. It was time to play again within the dreams of the living—their waking, sleeping, and interim dreams. He wanted to engage a real person, one who felt the heat of emotion and the thrall of desire. He had searched the roadways of memory for this woman and, recognizing the shaman's distinctive mark on her, had ended his search here.

He was looking at her now—the bright vision of a warm-blooded woman. Her face was pleasant to look at, and she was dressed according to her culture, in denim trousers and a white cotton shirt. She stood precariously at the edge of a cliff, with arms open, listening to the laughter of the swift hawks as they rode the mountain winds. The landscape around her was raw, sweeping, and wild. It was the finest terrain for an adventure—the perfect environment for a spiritual odyssey. Her odyssey had begun at birth, as everyone's does, as humans gulp air for the first time and wince against the glare of a new dream. Everything begins in that moment, but this seemed a good place to begin again.

The pleasure don Eziquio got from challenging other minds had begun when he was very young. It had started with a challenge to himself. Could he resist believing the words he spoke? Could he show his parents the respect they deserved and yet not believe them? Could he heed authority without submitting to it? He found he could do these things. He found he could use ideas as a child uses his hands at the beach, scooping wet sand into fanciful shapes, building impassable barriers in one moment and dissolving them in the next. A mind is a field whose soil can produce bounty or reject a harvest. Ideas are the seeds we plant, sometimes without intention; these seeds inevitably produce some fruit or another. If planted with intent, seeds call forth the sun and the rain to make a dream. The one who tends to his own harvest, aware of the process that gives life to a dream, is an artist.

Eziquio could see that Emma wished to be an artist. Perhaps she had been one in other ways, and through many avenues of expression. She had heard the word *toltec,* perhaps, and grasped its meaning. Her imagination thus ignited, she had decided to be the most intriguing sort of artist: an artist of life. Emma was not particularly young. In his day, she would have been considered past her prime, but her fire still burned brightly. She was inspiring to watch as she stood there, silently dreaming the dream of hawks. Did she know what had brought her here? Did she sense what forces had drawn her to this place, to this moment? The rock beneath her jutted out over a deep gorge, hundreds of feet above the Rio Grande. He saw her arms relax. He watched as she knelt, then carefully lay on her back, hanging her head over the precipice. Sky above her and river below, she would feel that she floated between worlds. *Are you ready?* he challenged silently, smiling. He liked this one. She was willing to push a little harder. She may not have known what she wanted—but he, don Eziquio, would guarantee that she got it.

As a student grows and evolves, so does the teacher. As the teacher's awareness expands, it serves as a portal for others, inviting them toward an infinite landscape. True, not everyone can read the invitation, Eziquio mused. Among those who do, few are eager to jump. She hardly knew what jumping meant, but Emma would not take the words literally. She had placed her body head-down on the edge of the cliff so that her mind might see the world differently; so she might sense the invitation. She was experimenting, playing, wondering. The old man recognized the strength of Miguel's investment in this woman's evolution. It showed in her actions, as she felt the silent urgings of his love and responded.

It felt as if he had been watching her for a long time before this. He had first noticed Emma in Peru, where she had obediently followed Miguel through mountains, meadows, and intemperate weather, exploring a dozen sacred sites in her eagerness to change and to learn. She was curious; there was no doubt. She had the courage of a desperate woman, but there seemed to be nothing desperate about her circumstances. Her time had come; that was all. This was her time to doubt, and to align with unknowable things. It was her time to slip unnoticed through the unassailable gates of reason and to run recklessly forward.

The attraction between teacher and student had been evident from the start. Emma's romance with Miguel had been inevitable. Further observation told don Eziquio that a lasting love seemed probable. Miguel lived within her heart. Eziquio knew what it was to win a woman's love and to earn her loyalty. He knew the precious consequences of trust and generosity.

His own life had encompassed several marriages and countless affairs. So much sweet music! So many songs, seductions, and betrayals! So much pleasure! He could swear that the old instincts were roaring within him now, thrumming up and down his spindly thighs and exploding in his loins. The feeling was real. The command was strong. It seemed impossible to escape life's fiercest ordinance. Every snail, every star, is guided by the same decree. *Be!* it says. *Be!* And so we become. Then we come together, so that more of life can be.

"Go to the heart," Eziquio muttered to himself, grinning. "Go to the heart." He had been called to save a life. What better way than to pluck playfully on a dying man's heartstrings? His great-grandson had relied on such genius, and he himself was offering it now.

The sun had set on the gorge cliffs, and Emma still lay upon the rock, her mind set free and her body exposed to the chill of evening. A dog ambled up to her and curled resolutely at her feet. He was an old dog, wolfish and grizzled, and he clearly belonged to her. Eziquio had seen this creature before, and he sensed that it wasn't she who had decided this, but the animal himself. He was hers. He was her confederate and the self-appointed guardian to her dreams, wherever dreaming took her. The old man suspected that the shaman breathed here as well, through the lungs of a hollow-eyed dog.

Don Eziquio watched the two of them from the shadow of a juniper tree. They were partners in a journey that neither understood, and hardly cared to understand. Life's genius was in the blood of all creatures. What remained simple instinct in the dog had to be relearned by the human. On a few occasions, Emma had lost herself in the high mountains, falling victim to summer storms and driving rain—only to have the dog lead her back to the old Jeep, and to safety. Together they conspired with life. Together, they followed the command of the *nagual*.

Ha! Eziquio exclaimed, as an owl hooted restlessly from the forest behind him. *Nagual*. The word came to him in the voice of his father, and his father's father. It carried the resonance of ancient times and sacred lessons. Emma was not a *nagual* woman yet. Challenging the mind sends a shudder through the human frame. Her world had broken in little ways and big ways, and she'd had the wisdom to leave its shattered pieces where they fell. She had begun to dissolve a lifetime of beliefs. Yes, she knew the lessons, she knew the course she was on, but for this kind of quest, even an enchanted dog would be of no help. He could lead her down a mountain, but she would have to find her way home . . . on her own. Eziquio smiled ruefully. She was an apprentice to a master, but where was that master now? Where was real Miguel, the one who had initiated the disturbance?

"Indeed!" a familiar voice boomed behind him. "Where is the heir to this wealth of lunacy?"

Eziquio turned on his heels. "Gandara!" he shouted.

There, in the soft light of evening, stood an old friend and fellow shaman. In life, Gandara had been a stout man with a stern expression, but death had changed him. He moved more slowly now, with none of his usual bob and bounce, but he seemed happier. His face beamed at the sight of his dear *compadre*. He snapped his suspenders against his chest and held both arms out in welcome, no doubt thinking that Eziquio had come to this place just to meet him. That was how it was between them, one always assuming superiority over the other. Gandara had taught the old man many tricks, but they were equals in skill and devilry.

In the old days, they were fellow conspirators, conjuring visions and terrorizing the countryside with their antics. They fed on the superstitions of local villagers and plagued the dreams of children. No obligation was more important than their plots and their pleasures. No job was sacred, no society exempt. When the fever for mischief was high, no woman was too virtuous and no enemy too vicious. Humanity was their playground, and they were the most fearless of players, both in love and in mayhem. They were legendary: *don Eziquio and his band of tricksters!* The old man was delighted to see his compatriot, his partner in crime, and his timeless friend.

"How in the devil's name did we come to stand on the same ground, you and I?" Gandara exclaimed, giving his friend a vigorous embrace.

"This is not my doing, *hombre*. I was summoned."

"For what purpose?" asked Gandara with heightened curiosity, eager to misbehave if called upon.

Eziquio considered for a moment, working the kinks out of his neck from the vigorous hug. "For the purposes of recapitulation," he offered.

"Recapitulation? We were done with that a century ago. I instructed you well, as I recall," Gandara said, slapping his friend on the back. "Extremely well."

"You instruct me? Are you drunk, miscreant?"

"I taught you many things, and well you know it. Before Gandara, could you steer your sleeping dreams? Could you read thoughts, spark emotions, conjure fire? Why, you hardly knew how to—"

"That's enough lying for one millennium, friend. My son Leonardo and I are retrieving the memories of another lifetime. We are attempting to recover a man."

"Recover a man! Are you speaking of resurrection, half-wit? Remember when we brought the unfortunate Pedrito back to life? His family was horrified that—"

"They begged us to do it! They insisted!"

"They were imbeciles—and so were we!" his friend bellowed. "Thank heaven he was taken by cholera so soon after. We would have paid a much heavier price!"

"Pedrito was beyond recovery, as we should have realized." Eziquio paused, taking a moment to shake the grotesque image of the resurrected Pedrito from his mind. "My great-grandson still lives," he concluded.

"Ah. Well, then. If he has lost nothing, then there is nothing to retrieve," said the stout man with relief. He hugged his friend again.

"One might say he has lost his will," Eziquio added, when he caught his breath. "My granddaughter cannot reach him, for all her power."

"Which one is this? Whose daughter is she?"

"She belongs to Leonardo," he mumbled, anticipating his friend's response.

"Ah! The singing anarchist! I remember him well!"

"He was a soldier," Eziquio corrected irritably.

"He was strumming and singing for the generals."

"*Gacho.* He was feeding his family and avoiding gunfire, just as you and I would have done," he countered, giving Gandara an affectionate slap on the belly. "We were built for misconduct, not combat."

"We were built to laugh, my brother. Have you forgotten?"

"Were you laughing when Nachito's good name was soiled—when he was banished to a life in the desert?"

He knew that mentioning Nacho would change the tone of the conversation. As a *compadre,* the man had been flawless. As a member of the brotherhood, he was the best of them. As a scoundrel, he was second to none. Eziquio watched his friend's countenance change, a cloud of regret dimming his joy.

"We agreed long ago not to speak of the incident," the man said cryptically.

"Long ago has come and gone."

Gandara, finding no argument, nodded his head. "Nachito," he mused solemnly. "He was accused wrongly."

"He adjusted to solitude admirably."

"He never implicated us, at least."

"We are an exclusive fraternity. We must always come to the aid of our own." Eziquio paused, mirroring his friend's grave expression.

"Ah, you remind me of our years of glory. How splendid we all were once—*don Gandara and his band of tricksters!*"

Eziquio glared at him but held his tongue. "We were legends, it is true; and we can be mischief-makers again, *viejo*."

"Bah! I'm done with the living."

"I sympathize. They have no—"

"No stamina. None at all."

"Exactly. Still, we have an opportunity to play," Eziquio said thoughtfully, "and to be generous, perhaps—in the name of our good friend."

"Nachito was without equal," Gandara reflected.

"An exemplary friend."

"How I wish he were here!"

"Why did he not join us?"

"He had his reasons," said Gandara.

"Such as?'

His friend looked at him soberly. "The fellow is long dead."

There was a prolonged moment of silence, followed by an eruption of laughter that made the mountains shake. The two men howled with joy, and howled some more, until the old dog was obliged to struggle to his feet, wag his bent tail, and utter a dubious *woof!* into the empty night. This stirred the woman from her reverie. She sat up, rubbed the chill from her arms, and stared into the looming darkness. The old ghosts, leaning on each other for support, looked over at her as they choked back their laughter. Eziquio tried to speak but instead began to cough uncontrollably. The dog barked again.

"Who is this?" Gandara asked, attempting to catch his breath as he wiped tears from his eyes. "A witch . . . with a wolf?"

"Hardly a witch, my friend," Eziquio said with a cough. "She is an apprentice to the new *nagual* master, clawing her way to transcendence with only herself as a teacher."

"Herself? You know better, you rascal," he said, jabbing Eziquio in the ribs. "She experiments with the elements, I see. Air is easy, thrilling. Earth is grounding." He paused, frowning. "Water was the most challenging for me."

"For me as well. I've watched her among the waves. Not here, but to the south. The more the ocean beats and churns her, the calmer she seems."

"Humans acknowledge the elements," Gandara intoned, "but only so much. They sense the danger of high places, so they cling to the rocks. They stay above Earth, fearing the dark. They avoid fire, knowing it will blister and destroy them. Few of them really know the elements."

Eziquio regarded the woman as she put on a jacket and collected her things. "She has spent her days and nights well: lured to lightning, drawn to wildness, dreaming in caves and animal burrows. She's danced naked in moonlight."

"A witch! I knew it! Tia Constanza was such a woman, remember? Every full moon, she was in the streets, running bare-bottomed!"

"Not a witch, this one—only a dreamer. We can show her how to dream farther, past knowing."

"Such a thing takes time. Do you imagine I have all the time in the world, *caballero*?"

"Yes, I do."

The stout man raised an eyebrow. "How does this help your family, old fool? How will it bring back the boy?"

Eziquio looked at his friend. Could he say what needed to be said, without being ridiculed? "One fire sparks another," he asserted, weighing his words carefully. "Fan this flame, let it rage, and the fire may spread."

"You speak of a man's love . . . for this woman?"

"For life, Gandara. And all that attracts life." He paused, unsure how his friend might be persuaded. "There is significance in that," he offered.

"Significance, eh?" The man's face brightened. "Nacho loved that word. Do you recall how—"

"Gandara," Eziquio interrupted, watching the woman brush dust from her jeans and call to the dog. "We may have all the time in the world, but still there is no time to waste."

"Correct," the big man said, nodding. "So be it, then." He cast a quick look in the direction of the woman. "She may lose her mind, you realize."

Eziquio heard the irony and winked. "This would not be the first time we threw old Mother Sanity on her back, *soldado*."

"And tickled her belly . . ."

"Bitten her massive breasts . . ."

"And shown her the fervor of *nagual* men!"

The two wraiths enjoyed another round of laughter, while behind them meteor showers exploded across the New Mexico sky. They could hear the old dog's distant barks of bewilderment, but they laughed anyway, kicking up bits of moonlight as they danced among the sagebrush. Thunder clapped from the horizon, echoing over the darkened plains like a celestial call to arms. *Bring on the bedlam!* it cried. *Something new is in the air! Time, bored with pretense, is keen to betray itself!*

Eventually, the skies calmed. The valley settled. Jubilance was tempered as the planet rolled into a restless sleep at last. The two old men resumed their conversation in hushed tones, and the night listened quietly. Stars were gleaming modestly when the men took leave of the high bluffs and strolled away together, shoulder to shoulder, scheming softly in the shadow of the mountains called Sangre de Cristo.

◉

Great-grandfather, the wizard of my childhood imagination, I see you now. I see the sprite, dancing among a starscape of thoughts, gleefully rearranging galaxies. I see the imp, toying with dreams. I see the hellion, racing with a bucket of magic to douse the fires of hell. In you, don Eziquio, I see myself—always playing, always eager to leap into the dark. You were right to seek the spirit of the teacher within the heart of his pupil. I've lived within Emma's heart for a long time, as she still lives in mine. A master will inevitably open channels between himself and a good apprentice, channels that are seldom available to others. I left my shaman's mark on her many years ago—a psychic imprint intended to defend her from any sort of harm—and it's no surprise that you followed its traces here, to my beloved, and into her deepest dreams.

I feel a natural connection to my great-grandfather, although we have no common memories. Dreaming of him gives me a strange comfort, and it reminds me of life's happier absurdities. Dreaming, asleep or awake, is all the mind does. What seems real while we sleep looks different when we awaken, because our waking dreams adhere to strict rules and probabilities. Awake, we dream like everyone else, agreeing on what things mean and submitting to common assumptions. As awareness expands, dreamers can see more, and reveal more to themselves, than they normally would. With attention, and a willingness to change, apprentices can return home from a power journey and see their life differently. They can respond to the people in their dream differently, and change the reality they share with those people—for the better.

Taking a power journey is, at its essence, an opportunity to experience another way of perceiving. A glimpse of truth has a lasting effect, even if it becomes modified over time. Everyone journeys differently, but a slow process generally works best. This is the case even for the many students who demand that I push them harder and hold nothing back. "Have no mercy," many say. The most rigid minds are the most insistent, it seems, but there is no freedom for them—for us—unless we first liberate ourselves from our own judgments and assumptions. There's no progress until we're able to listen to ourselves. Every statement of fact should be questioned. Every conviction is a possible blind spot. Life becomes so much simpler when we stop defining things as good or bad, right or wrong. We assert our self-importance in many ways, and we cause a great deal of pain as a result. I've given my students titles, pet names, and positions of authority. Eventually their pride is exposed, providing an invaluable lesson for them. It's not importance that allows us to transcend ourselves; neither is it piety or humility. And it's not knowledge, secret or otherwise. Transformation happens to those who change the world—the world they create in their own minds. Many people are willing. A smaller number really *do* change . . . and some never stop changing.

Emma entered my dream long before she became my apprentice. I first met her in New Mexico, during a speaking engagement there. She was a seeker without a community, and without support from those close to her. She saw that she had lost someone she loved: herself. She was unable to recognize the person she had become, and incapable of resurrecting the person she'd once been. The truth was hard to see, but she had the potential to reach it. She needed guidance. How do we change, when we can't imagine the possibility of change? Seeking authenticity—something society has failed to cultivate—we feel like confused children again, without parents this time, and without instruction. Emma was mature at this juncture, married, and trying to dig herself out from the rubble of lies. She wanted truth but could hear only her own beliefs and opinions. In that way, she was like any one of my apprentices—almost. She may not have known at the time of our first meeting, or even in the initial years thereafter, but she was a seeker of a different kind. She was a warrior for change, ready to see, to forgive herself, and to act . . . and it appeared she would never stop.

E MMA AND HER SIX-YEAR-OLD SON SAT ON A
mound of dirt by the side of the road. She found one
stick of gum in her pocket and split it with him as
they talked quietly about the shape of clouds, the warm sun,
and the beauty of this January morning. They had been on
their way to the hot springs, both feeling naughty for her son's
skipping school, when her car skidded on gravel and flipped
completely over. They were both shaken but uninjured. They
were calm, even happy, as they chewed gum and waited, not
knowing what might happen next. It was unlikely that anyone would come by to help before day's end, if then. Although
this stretch of road eventually led to other, bigger roads, it was
rarely traveled. Emma looked wistfully at the car sitting in
the sagebrush. It was settled comfortably on four flat tires but
crushed almost beyond recognition. The roof was caved in, the

hood crumpled, and all that remained of the windshield was a few shards of shattered glass.

As her son chattered about school and about his friends, Emma calculated the number of times something like this had happened to her in the last year. Maybe four; this would make five. Car crashes, a spinout, a sideslip, and an incident when the car went sailing through sunlit space and landed in a canyon of snow. Once, just months before, she had slipped off an icy country road and hit a barbed-wire fence. The front of the car had ended up wrapped in spiked ribbon like an early Christmas present; in fact, some of the twisted wire was still lodged in the headlights today. Another time, coming back from Colorado, she was driving alone at a high speed when her car swerved and started spinning uncontrollably. She was growing accustomed to this feeling, this loss of control, this . . . availability to death. After so many incidents like this, however, there seemed little chance that death wanted her.

Since her first trip to Teotihuacan, she had thought of death as a necessary beginning. It would lead to transformation, to truth. Transformation was a much more perilous affair than she'd anticipated, and truth obviously lay beyond her small powers of understanding. Since meeting don Miguel, she'd been hurled into a different realm. All paths led inward and all roads appeared treacherous. Here, nothing made sense. Here, love was no longer an emotion, but the sum of all emotions. Here, the weather changed from minute to minute, and typhoons of feeling were frequent. She cried for no reason and laughed inexplicably, as if the two were the same. She'd become secretive, silent, gathering insights like a madwoman by a midnight bonfire. Her waking moments were filled with bright new revelations. Her nighttime dreams were alive with characters—old crones; villains; and faceless, genial guides.

Snakes leaped through her visions, hissing and striking, biting her face, hands, and legs. Crocodiles talked to her, offering advice and polite warnings. Ancient symbols found new life in her. In her dreams, she planned voyages, packed bags, and missed trains. All was chaos, awake or asleep.

She didn't know how long this disorientation would last, or if it was ever supposed to end. It seemed that once she'd signed up for change, the chaos was unavoidable, and she'd have to settle in for the long ride. Abnormalities were normal now. She was tired of wrecking her car, tired of stumbling and babbling, but it seemed she must learn how to move again, and how to speak. In moments of pure wonder—and the wonders kept coming—she gave up knowing things.

Emma sighed and tossed the gum away. Her son was pointing to a rabbit peeking out from behind a clump of chamisa. He leaped across the dusty road to get a closer look, inadvertently chasing the bunny out of sight. A golden eagle rose out of the brush nearby. Some would have called that a good omen. Today, she would agree. It was a good day, no matter how things unfolded. Taking another look at the car, however, she began to think her friends were right to worry. Inviting disturbance was one thing. Risking her son's safety was another. This, at last, was too much.

"She is right, *hombre,*" Eziquio agreed. "This is too much."

"This was not my doing, friend," Gandara objected. "This was the work of other entities—nefarious, contemptible entities. You know how it goes."

"That story might have worked when I was alive," his companion said scornfully, "but we are on the other side now, Gandara. Look around you. What entities do you see?"

"Just you and me," he conceded, shrugging.

"Just you and me, and we are no more than figments of a dream."

Gandara seemed to take offense. He slapped his belly vehemently with both hands. "Does this look like a dream, *patrón*? Feel it!" He reached for the other man's hands, but Eziquio walked away. The road stretched southward, toward piñon hills, and Eziquio headed in that direction. An eagle, calling, swept across their path. "Feel the dream, coward!" Gandara shouted again, running after his friend.

"There are no nefarious entities, old man—just you," Eziquio said when his friend caught up to him. "And you are an idea . . . no more dangerous than that."

"And no less so," said Gandara breathlessly. "No less so."

There was no contradicting that, Eziquio thought. He huffed and said no more. He wanted their venture to work. He wanted Miguel to feel the woman's anguish and be moved again. He wanted passions to burn and love to jolt the human spirit back to life. It was only reasonable that Emma would die a few significant deaths in the process—but it was not reasonable for her *body* to die. He fluffed the faded ruffles at his wrists and strutted on, tight-lipped and peevish. Gandara, panting next to him, seemed repentant.

"Well, *compadre,* I tried," he offered.

"Warriors die trying," Eziquio snapped, and then was silent. He slipped deeper into contemplation as he walked. He had wanted to fan the flames, of course. By raking these embers, by reliving and intensifying Emma's fervor, he'd hoped to stir the waning heartbeat of another. It still might work, he thought. Impassioned dreamers touched all worlds. What harm came of nudging a domesticated mind farther into the wild, if its screams might ripple through time? Emma was in crisis. She was flitting from place to place, from story to story, from the familiar to the strange. Real and unreal were becoming confused. She may have been fighting alone, but she was calling out for the shaman with every breath.

Was he hearing? Was he remembering the vibrant connection between them?

The midwinter sun was hot, but cool winds rolled down from snowy peaks, carrying drifts of pine-scented air. Time lazed along; no man could tell how fast or how slow. As shadows inched across the plain, the two men walked on. The chill grew as they approached the mouth of a canyon. Hesitating, Eziquio removed his hat and turned his weathered face back toward the mountains. How exquisite life was on this welcoming planet! This dalliance, this brief, imagined existence, summoned in him such feelings of desire!

Desire begins everything, he remarked to himself; it comes before every imagining, every action. Desire sets life in motion. Desire is the spark that ignites the flame—desire for a woman, for a dream, for existence. There was a time when life felt the first stirrings of desire and, in response to that feeling, created a vision of itself. What life desired was instantly created; and the creation desired life in return. Desire spawns desire. Every dream is an integral part of the thing that dreams it.

"We live and breathe within life's dream," Eziquio uttered aloud.

"You stated earlier that you and I exist in the dreams of others."

"How is that different? We are all the consequence of life's libidinous desires."

"Ah! I see." Gandara nodded. "Well put."

Eziquio grinned, dry lips cracking, as he placed his hat firmly on his head. There was a time when such a revelation would have taken weeks of preparation and barrels of wine. Now, from his present vantage point, clear comprehension required hardly any effort and caused not a ripple of disturbance. He understood the woman sitting by the side of the road. He understood her yearnings in ways she could not. Living by the rule of the *nagual* master, she was now his quarry, targeted for disturbance. She belonged to him, whether she was near him or far away. She felt his taunting, his coaxing, and his incontestable love. It was still possible that, from the distant corridors of this memory, Miguel could feel hers.

Eziquio, now seeing things in a better light, ambled lightly down the dirt road. He was pleased with his part of the mission, in all events. Two old friends had met again to play among the living and feel life's joyful complicity. He was pleased with the dream, proud of his great-grandson. Yes, the boy was an Eagle Knight, he thought—to the bone! Miguel had fueled the ancestral magic, allowing these two old men to meddle again. Clearly, his apprentices delighted him, far more than even Miguel remembered. Their willingness to play was a wonder; their courage in the face of shifting realities was inspiring. Why, Emma was—

Eziquio felt a pang of misgiving and stopped in his tracks. "She and the little boy must find a way back home," he said.

Gandara halted, gasped for breath, and then looked back, considering. "No, they must find a way forward, *jefe*. A dreamer doesn't let misfortune impede the fun."

"And how, precisely, should they go forward?"

Gandara, feeling the urgency of the moment, wiped his sweaty face with the palm of his hand and concentrated. "Ha! Do you remember Nachito's son? The one with the rusty blue bucket?"

"Bucket?" Eziquio asked. "What are you talking about?"

"*Si!* The truck that never stopped coughing!"

"Ah! The monstrous bucket!"

"The village was horrified by the sight of it," Gandara went on, "as if it were a creature out of fantasy—a dragon baring fangs and spitting smoke!"

"A truck? That's your answer? Did we need trucks when we held Veracruz? When we marched on the capital? When we lifted the Freemason to his glory?"

"Gods. *Por favor,* no more talk of General Juarez. You know it makes my feet bleed just to hear his name." Gandara was breathing hard and frowning, as his mind worked to solve the immediate problem.

"Then get the woman home!" Eziquio commanded. At that moment they heard the cough and sputter of a machine behind them. Both men turned in surprise and saw a faded blue truck bouncing in their direction, looking every bit like the wreck Nachito's son had driven through his village almost eighty years before. A little man sat behind the wheel, accompanied by a towering dog in the passenger seat. He lifted his hand in greeting as the junk heap drove by, leaving both men choking in a cloud of dust. There it was. The boy and his mother would be found, and their carefree day could continue.

Eziquio wanted to reassure her, to tell her that once the world was fully tipped, everything would make delirious sense—but he doubted she would believe him. For a while, she would continue to doubt. She would resist solace and mistrust simplicity. She was not prepared to let go of the struggle. A few more heartbreaks, several more little deaths, and she would find balance. Truth was ruthless, but she would meet it. Life was joyful in its antics, but she was resilient. They had lured her to the edge of reason, but reason would expand its present boundaries, as it always must. He himself had played tricks on her—many marvelous tricks—but what was don Eziquio, if not a trick of light? It struck him how easy things were when one was no longer bound to matter. Yet, even within matter, freedom could be experienced. When life had still coursed through his veins, he'd been awe-inspiring. Extraordinary powers had belonged to him then, and to Nachito—and yes, to Gandara as well. He glanced at his friend admiringly and patted him on the back as they set out toward the foothills. Even fat, lumbering Gandara had shown the reckless skill of a sorcerer in his day. Long ago, the three of them had mastered the art of leaving the world of "real things"—and coming back. Coming back was the hardest trick of all.

Again he thought of Emma, confronting the same fears that all people have confronted since the earliest days of humanity. Their fears always came disguised as reasonable questions: "If I change, how will I know myself? If I stop believing, will I go mad? If I travel too far, can I find my way home?" When one is aware of oneself as everything, home is a fool's delusion. Eziquio chuckled to himself. No boundaries existed in life's perception. As a younger man, he'd been able to skip effortlessly between dreams. He'd had a knack for lifting off, and for landing again safely. He'd felt no curiosity about it all then, only joy, and the sheer thrill of leaving. Leaving, and coming back. He'd had what most humans never dared wish for. He'd had the audacity to jump out of knowledge.

And what of don Miguel, the heir to his legacy? Where was he in this landscape of memories? While his ancestors played, while his apprentices balked and trembled—while love called and heard no answer—what was the man up to?

"Gandara," he said suddenly, "I want to see the teacher in action."

"The Nazarene?" Gandara asked, then quickly apologized. "*Disculpa*. I lost myself in time," he muttered.

"Miguel is up to something," his friend deduced. "Before this moment or after it, he has been up to something!"

"Shall I find him?" Gandara asked halfheartedly. Nothing about this adventure had seemed easy. Even in this moment of peace, for instance, the sun was getting hotter and the road steeper.

"Yes, find him," Eziquio answered, "even if it means jiggling time."

"And shall I also find you a strong drink, *patrón,* and a hot meal?" Gandara added. "What about a hot woman—perhaps you would like to jiggle that?"

"I admire your work, *amigo,*" said his companion, sensing that he had offended the old man. "Your piece with the truck was masterful. Nachito would have given it a dancing ovation."

"Yes, yes. Nachito would have been amused," his friend agreed, "but back to the point. While I am jiggling time, what of you?"

Eziquio swept off his sombrero again and took in the exquisite heat of the sun. He puffed his chest out, simulating a deep breath, and exhaled audibly. "I will be here, resting on my laurels, *vaquero,*" he sighed theatrically. "On each of the many laurels of my long and meritorious life."

Gandara heaved a sigh, shaking his head at the fool who stood before him—a character for all time, but a loyal kinsman beyond everything. He must heed the wishes of his friend. After all, once desire was kindled. . . .

Before he could reflect on the power of one man's desire, time altered to life's will and the brilliant sun was no more.

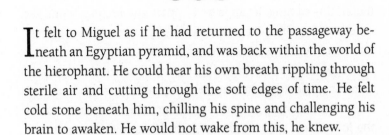

It felt to Miguel as if he had returned to the passageway beneath an Egyptian pyramid, and was back within the world of the hierophant. He could hear his own breath rippling through sterile air and cutting through the soft edges of time. He felt cold stone beneath him, chilling his spine and challenging his brain to awaken. He would not wake from this, he knew.

He opened his eyes and scanned the familiar space. Torches lit the hallway, throwing restless shadows across his face and crimson flames along its walls. These were the same high walls, sculpted with hieroglyphs from ceiling to floor, that he had seen as a younger man. The feeling was the same: the faint scent of spices again reached his nostrils, and the taste of minerals touched his tongue. The very moment seemed the same. He tried to lift his head but could not. His body was lifeless, helpless. His mind reeled, trying to make sense of this dream again, and finding no appeasement. Had he missed something before? Had he failed to understand? Was he here to reinterpret the lesson? Unable to move, he lay still, waiting for the robed hierophant to enter the chamber.

No one came; nothing happened. The torches flared and spat, but there was no other sound now. Miguel couldn't be sure if the hours flew quickly or grew stagnant. If time had come to a halt, he didn't know. He lay there, running his gaze over the stone engravings above him, until one larger image caught his eye. It had been painstakingly etched upon one of the walls, and it depicted a story he knew. The characters were drawn differently from those he was accustomed to, arranged differently, but he knew them, too. He couldn't name the gods and goddesses, but he recognized the principal actors in this story. There were two, in particular. One was the lord of the underworld, Guardian of the Scales, who weighed the hearts of the dead to determine if they were worthy of the eternal afterlife. In this etching, he appeared as part human, part wolf. The other was the Great Devourer, colorfully represented as a combination of the largest meat-eaters known to humans: a lion, a hippopotamus, and a crocodile. Upon the enormous scale, the Guardian balanced a human heart against an ostrich feather. A heart that was heavier than the feather would be devoured by the formidable creature. If it weighed less than the feather, it would be allowed to pass through to the eternal realm.

What determined the purity of that heart, of any heart?

Miguel looked at the symbols in the drawing, so brilliantly imagined and offering such a simple lesson in transformation. He had spent a lifetime cleansing his mind of poison. He had searched his heart, found the corruption, and removed it. He had torn away the lies—lies he'd been told and had believed, and lies he had devised and used against himself. This was his mastery—the purification of the mind and the recovery of authenticity. This scene symbolized the mastery of death, the awakening.

Miguel looked again at the figures, surrounded by all the intricate symbols that told the story. He looked at the bodies of animals and humans—their dress, their masks, and their instruments, all drawn in muted tones and delicate shapes. The penalties of neglect were meant to seem scary in this story. Whether we are kings or slaves, it warned, it is up to us to remain the vigilant guardians of truth. We are the corrupters and the victims of our own corruption. We are the gods who make choices concerning our own freedom. We are the Guardians of the Scales, presiding over the ritual, as we weigh lies against the truth of us.

Suddenly the images faded; Miguel could barely make them out now. Perhaps he was dreaming after all, and this dream was coming to an end. Perhaps eternity was waiting for him just beyond the flickering torchlights, just beyond the darkened corners of the room. Yes, eternity *was* waiting. It felt close . . . as if it lay between each carved symbol and underneath every fleck of paint. It was close, waiting, and eager to welcome him. Now the smell of spices was fading, too.

Feeling his senses grow dull, Miguel closed his eyes and surrendered to a deeper sleep. His breath softened and his pulse slowed gradually, ever so imperceptibly, to a stop. All was well. Eternity was waiting.

"Is he dead?" There was a hush within the chamber, but

those words, so softly spoken, echoed from its granite walls in ghastly parody. *Is he dead? Is he dead? Is he dead . . . dead . . . dead?*

The woman kneeling beside Miguel's body held his wrist, trying to find a pulse. Sensing nothing, she leaned her left ear close to his mouth. No breath. One of the men knelt beside her and went through the same motions; then he placed his head against Miguel's heart. After a full minute, he sat up and looked at the others. *No heartbeat,* his expression said. The shock was immediate.

"No!" someone gasped, and the echoes gabbled around the chamber again. *No! No . . . no . . . no!* The entire group of students reacted this time, talking, questioning, shouting, all at once. Several more crouched near his body, touching him, calling him, and even shaking him to wake him up. Women began to weep, and men to wonder helplessly.

This was the second day of their group journey to Egypt. They had come to the Great Pyramid that morning, filled with excitement and eager for a private tour of the world's most secret wonder. The tomb of Pharaoh Khufu was situated high in the interior of the Great Pyramid of Giza, in the so-called King's Chamber. It was a large, rectangular room faced in polished granite, containing nothing but a sarcophagus. This lacked the grandeur of a king's coffin, but there was no denying that its presence suggested kingly mysteries . . . and unnamed peril. The Queen's Chamber, placed well beneath this one, almost in the middle of the great structure, felt more regal and far less ominous. It had seemed a welcoming place.

The group had felt as if their arrival had been warmly anticipated from the time they'd gathered at the base of the pyramid, entered through the broken outer walls, and begun their solemn march along its gleaming passageways. Their awestruck whispers lifted into the unseen tunnels and reverberated through every chamber. Reverence was met with recognition, or so they felt. Their cheerful mood seemed to rush ahead to change the place, change the very tenor of history. The pyramid was closed to regular tourist traffic that morning, allowing them to explore its marvels without distraction and at their own pace. No one was sure how that had happened, but it was accepted as fact that *everything* was possible in don Miguel's world. Why ask why? They had sole access to the Great Pyramid today, and today such things were right and normal.

Suddenly, though, as Miguel led them into Pharaoh Khufu's chamber, nothing felt right. Nothing was as it should be. What had taken place to change things? they wondered. What had they missed? As everyone crowded into the chamber, finding a place to sit or stand while they took in its moods, Miguel walked its perimeter, brushing closely against its high walls, hands behind his back, and inspected the high ceiling. It seemed as if he knew the place, or someplace like it, and was in the process of remembering. To some, it seemed as if he were greeting old acquaintances. Having walked the room, he went up to the sarcophagus and stood beside it for a while. Then, without a word, he lay down on the floor, closed his eyes, and placed his arms across his chest. His apprentices, accustomed to the ritual of dreaming, closed their eyes with him . . . expecting nothing, but waiting.

They waited . . . and waited. Hearing no sound and observing no movement, one woman finally knelt by his side and whispered in his ear. Getting no response, she took his pulse. . . .

"There are no sure ways to tell if a man is truly dead," Gandara stated, watching from the back of the room. His words, however clearly spoken, made no echo and caused no reaction in the crowd. "Believe me," he added, shaking his head.

"Believe you?" Eziquio was right behind him. "You almost buried me alive all those decades ago!"

"I knew you were still alive," his friend blustered. "I knew. Why else would I postpone the funeral?"

"You did nothing of the sort! You left me on the parish steps for Padre Quique to find me and to bury me! You ran from death like a frightened girl!"

"Ridiculous!"

"Yes, you are!" Eziquio said, raising his voice to make himself heard as the crowd around Miguel began to react to the apparent demise of their teacher. "Just as these people are being ridiculous," he added.

"Ah, these people!" Gandara said, happy to change the subject. "They lack the temperament for wonder. They lack—"

"Patience," Eziquio said flatly. "They lack patience. Miguel's mother would have waited. Sarita would have gone into her kitchen to make *posole* for everyone, and she would have left him to his dreams."

"I had such a mother," Gandara intoned. "A shaman's best friend." He turned to Eziquio and smiled. "Felia, my wife, would never let me do things like this. I had no freedom to play. 'Who will buy the corn and the chilies?' she would nag me. 'Who will carry water from the well?' she would insist. 'Who will provide for us, with you falling dead all the time?' So many question, objections! Mother detested her."

The two men observed the scene silently, as people circled closer to the body, fussing and whispering, their initial outburst having abated. Their Egyptian guide had gone for help, and it appeared little else could be done. Eventually, even the whispering receded, leaving only the sound of choked sobs and stifled tears.

"Wait!" Gandara said abruptly, giving his companion a suspicious look. "I thought we agreed that you would rest on your laurels."

"I agreed to nothing. I wished to find the *nagual* teacher, and I have." He pointed to the scene in the chamber, then looked his confused friend in the eye. "What he does affects them all. What he dreams affects everything."

"I can handle this on my own, *viejo*," his friend assured him with a nudge to the ribs. "Go back to the mountains and tip the world into madness. Go. Have fun."

"The world can wait," Eziquio said. "I want to see what happens here."

"He dies, or he does not. What else?"

"Significance, *hombre*. I want to determine the significance of the moment."

"Must we seek hidden meaning in these things?" Gandara grumbled. "We live. We laugh. Why should there be meaning?"

"There is no meaning to it," Eziquio snapped. "*Significance* is another matter."

"He will pay consequences for this, if that is significant."

"The body pays a price for dying, as I well recall."

"Nacho left you on the chapel steps, *amigo*. Not I."

"Ha! There it is!" Eziquio turned to his friend, a spark of malice in his eyes.

Just then, a woman gasped. Someone shouted, and the crowd backed away. The two old men, peering through the moving mass of people, saw Miguel, eyes opening, slowly sit up. Three women rushed to his side, helping him to stand and speaking to him reassuringly.

"Good work, *m'ijo*," Eziquio muttered under his breath.

"Look at his face," exclaimed Gandara. "He has no understanding of what transpired. Can they be sure he is alive?"

"I suppose they can place him in the ground and throw dirt on his face."

"That was Nachito, I assure you."

"Liar! I awoke and saw you—"

"Hush!"

Miguel, now upright, was calming his students and gesturing for them to be seated. Everyone in the chamber became silent. He stood before them, one hand resting lightly on the stone sarcophagus, and said nothing for several minutes. The large room seemed to close around them, muffling every sigh and subtle movement. Shimmering walls illuminated their expectant faces, and the pyramid itself seemed to catch its breath, settling in to listen. Standing before the group, his black eyes shining, his voice soft and caressing, don Miguel began to talk.

"Put attention on my words," he began. "Don't be distracted by your stories. Don't believe what you think you know, or what others want to tell you about this moment."

It seemed as if his voice came from the belly of Earth, low and resonant, shuddering through granite and flowing through the main arteries of the pyramid, its massive body humming in response. Above, hidden compartments murmured. Outside the room, memories whispered. Gold threads laced the granite walls, and the burial chamber of the pharaoh seemed to reclaim its regal provenance.

"I want you to know that I will always be here for you," he said. "As long as you live, as long as you can listen, I am here with you. I will come back. I will return again . . . and again, to deliver the message of life."

"There it is," Eziquio whispered. Jump out of knowledge, he thought to himself. That was the rule . . . and now it was done.

Miguel continued to speak to each mind and to every open channel, although it appeared that he spoke to nothing and to no one. His eyes were bright, but what he saw was beyond them. He was looking through the smooth granite walls, perceiving beyond the world of matter, and gliding above the tremor of this fragile moment.

I remember the sensations of surrendering to a long-abandoned dream and sailing instead past all dreaming. It brings me a quiet serenity to be there again and to see the experience from all points of view. My trip to Egypt was precipitated by some difficult events, including another long journey to Peru. Before the trip, my sleeping hours were touched by ominous dreams, with images of old gods and demons, symbols of ancient Egypt. It seemed my mind was reflecting on a collective dream that once was, one that had faded and disappeared as humanity's attention moved on. It was not my intention to revive that dream, held deep within the vault of human memory. Yet even simply by going to Egypt, I felt I would be summoning elements of old mythologies that might challenge me. To be cautious, I expressed a wish that the people closest to me stay at home. I told Emma not to come with me; I advised her to keep as much distance between us as she could, both physically and emotionally.

This was the way reality was structured in my imagination then, and everyone around me was a part of that reality. My apprentices were in my dream, by their own volition, and the rules of reality changed accordingly. Within the domain of my dream, people assumed the language and the logic of don Miguel. I used shamanic symbols, and so did they. They spoke a language built on those symbols. They lived and dreamed those symbols with me. As my teachings changed, I changed symbols to suit my shifting approach to awareness. Those who responded well to change stayed close to me—and, with some big shifts in the dream, the teaching grew. Those who disliked the changes fell away.

All shamans are different, but their apprentices are swept

into the shamanic dream in similar ways. Tucked tenderly under his wing, students are given the benefits of a teacher's attention and opportunity for growth. In my dream, respect is everything—respect for oneself, for others, and for all of creation. Mine is a dream that discourages fear and creates an atmosphere of unconditional love. That has been the mood of every journey, including the one to Egypt. Whatever anxieties may exist, my students resolve them and go forward in awareness.

Once I arrived with my group in Cairo, everything went smoothly. I was given permission to take the group on a private tour of the Great Pyramid of Giza, and from there the journey would continue down the Nile. From the initial moments inside the pyramid, I felt comforted, relieved. The massive structure enveloped us like a womb, and the dream of an ancient people became intimate to me, familiar. In the King's Chamber, I experienced something like my dream state with the hierophant from years ago. I began to understand the place, to understand the human drama as it existed in ancient times. I lay down on the floor of the chamber and surrendered. At that point, it was as if I expanded into the dream of life itself.

When we refer to body, mind, and spirit, we create misleading distinctions. There is only life, and life's countless points of view. Detaching from one place, one circumstance, and one moment, I was no longer bound by my own perceptions. I *was*—and that's all. What does it mean to see from life's point of view? What does it mean to be infinite in the present moment? It may be impossible to say, but it is possible to experience. People say I was dead in that chamber, but that idea belongs to the brain, to matter. Dead? Gone? Out of body? These notions, like all notions, have no relevance from life's perspective. In many ways this incident was a prelude to my eventual transition. I was not dead then, and I am not dead now, but my curiosity about death has been satisfied. On the journey from infinite potential into the dream of matter and form, and then back to infinite potential, there is no disturbance. From life's perspective, nothing happens.

From the perspective of matter, however, there's a drastic disturbance. Matter is short-lived and affected by the upheavals of change. Regardless of the person, regardless of the species, we are all dreaming reality from the point of view of matter, even if we have the awareness to see ourselves as more. Our ideas of life and death—and everything else—exist within the dream of matter.

Matter has a powerful memory, something we know because we have a brain. The brain remembers, whether the mind wants to or not. Memory has an enormous influence on the way we perceive and what we believe, but it is just one function of an organic mechanism whose abilities are incalculable. The senses, language, reason—these are all functions of matter. So are emotion, mood, instinct, and intuition. I want to emphasize the complex nature of the physical body, and to explain that our perception is only as evolved as matter itself. The evolution of matter is measured by the way it responds to light. Human evolution has everything to do with the changing complexity of the human nervous system and its sensitivity to light. Light is life's messenger and its first manifestation. This may sound like story, and that's because it is. I'm putting words to something that cannot be accurately explained, as anyone must do who searches for the truth through symbols. Many enduring stories have been told about light and about life's messengers, but we are likely to attach to the story instead of its underlying truth.

Let's say that life is an unknowable entity, a nameless force. Let's say that before matter existed there was only the potential for existence. Let's say that potential, or absolute power, is suddenly compelled to perceive, to look. Looking creates two points of view—the seer and the seen. The mechanism connecting two points of view is light. Like a child peeking through a keyhole, life looks around, and light reveals a universe. There is no difference between the central characters of this story. There is no distinction between life, what it sees, and the means by which it sees. Pure potential has expanded into pure perception. The "keyholes" are everywhere, and the views are the same. Life is seeing, and life is being seen.

There was nothing strange about my experience in the Egyptian pyramid. I dreamed, and then I was dreamless, timeless, and eternal. I was not speaking as Miguel when I told my students I would return, but as the pure power of life. Life is always there, offering a new beginning and another opportunity for discourse. My heart was light. It was weightless, no longer suffering the burdens of knowledge and the tyrannies of the storyteller. Life's message was clear to me as I slipped past death that day. I felt no threat, no sense of things ending and beginning. In a way, that day in the King's Chamber was a portent, pointing to my final shift in awareness.

That shift came two years later in Teotihuacan, when I realized that my physical body could not endure me much longer. It would not long survive the force of life, ever expanding and intensifying within it. A week later my very real heart would fail, and one way or another, the Miguel that I had known—and the Miguel that everyone else had imagined—would die.

19

I WANT TO PLAY!" JOSÉ YELLED AT HIS BROTHER.

"Forget it," Miguelito said sharply. "*I'm* playing."

"The game is mine, too!" José tried to grab the controls from his brother's hands. At twelve years old, Miguelito was bigger, but José had managed to wrestle him to the floor one time. He was ready to do it again.

"It's my Christmas present," his older brother snapped. "And I'm playing."

"Come on! Let me be Luigi!"

"I want to play, too!" cried little Leo, joining the uproar.

"Hey, what's going on?" their father asked as he walked into Maria's house to pick up his sons. "I could hear you from the street."

"Pops! I want to play!"

"You can't!" yelled his brother.

"Play what?" asked Miguel.

"Super Mario, Dad."

"Nintendo! And it was a present for both of us!"

"Why can't he play?" asked their father.

"Because I'm playing."

"I can play! We can both play!"

"Is that true?" Miguel asked.

"Yes!" said José.

"It's not on Luigi mode, Luigi," groaned his brother.

"See, Pops? I'm Luigi—he called me that! He should let me play!"

"I want to play!" echoed Leo.

"What's the game about?" asked their father, sitting beside his oldest son.

"I can't explain it," Miguelito moaned. "It's too complicated."

"It's simple!" José contradicted, throwing himself onto his father's lap. "Mario and Luigi go into the Kingdom of Mushrooms to save Princess Old School."

"Toadstool."

"They have to save the princess, but the bad guy is trying to kill them."

"Me," his older brother said. "He's trying to kill *me*."

"Me, too, if I can play."

"So what do you have to do?" asked Miguel, staring at the screen.

"I have to go from here to here—see this flagpole?—without getting killed."

"I can help!" José insisted.

"Why can't he help? He's your brother."

"He's not any good at it."

"I am too!"

"Me too!" said Leo, crawling over the back of the couch to peer at the game in progress. His little hand reached for the control and was slapped away.

"Do you have weapons?" asked their father. "Armies? Strategies?"

"He's alone," said José, "but I could—"

"See these little blocks with question marks?" said his brother, pointing to the screen. "They hold stuff that I could use to defeat the—"

"They have coins and cool things," José interrupted excitedly. "There's also the mushrooms! Show him!"

"This mushroom makes me grow bigger and stronger."

"Some give you immortality!"

"*Lives*, stupid," his brother said, rolling his eyes. "I get more *lives*. See? He doesn't get it, Dad."

"Don't call your brother stupid."

"He is. They both are."

"I'm not," said little Leo, hanging on his father's shoulder.

"Show me how it works, guys," said their father calmly.

"Let me, Pops!" José volunteered, bouncing with enthusiasm.

"Let me!" mimicked Leo.

"I'll do it." Miguelito groaned again. "Dad, there's eight worlds in this game, and a bunch of stages in each world, and I have to move through them all to win. Only it's really, really hard."

"Not for me!" exclaimed José.

"In the last stage of every level—"

"Every world," corrected his brother, "stupid."

"In the last stage, I have to fight this Bowser guy, or some friend of his. See how big these guys are?"

"I know how to fight him!"

"Now, if I can find a warp pipe—"

"Or Starman!"

"—I can skip some of these levels," Miguelito continued. "I mean *worlds*. But it's hard, and I keep getting crushed by these guys. Once I drowned in one of the underwater things."

"Yeah," said José, his brow furrowed.

"Wow," said Miguel, his face brightening. "Do you see how this is just like life?"

"Not my life, thanks," said his oldest son. "Except for having a stupid little brother named Luigi."

"José Luis, stupid."

"Stop," their dad said firmly. "This is like everyone's life. We're born into a game we don't know how to play."

"A game?"

"Human existence. Yes, it's a game. It has a lot of rules, like this one. Have you forgotten all the rules you've had to learn since you were born? Didn't you have to learn how to go potty by yourself?"

"Potty!" Leo rolled off the couch and hit the floor laughing. "Potty!"

"You had to do what your parents said," Miguel went on, "and what your teachers taught you, right? You do a lot better in human life when you learn the rules, when you learn about good manners and fairness. You do better when you brush your teeth, go to sleep on time, and help your *mamá*. You do better when you get an education and become skilled at something."

"Well," said Miguelito. "This game lets me live again after I've been killed. It gives me superpowers. If I pick the Fire Flower—"

"The Fire Flower!" cried José. "Boom!"

"Okay," said their father. "There are useful strategies in life, too. When you're aware, you see things no one else can see. When you believe in yourself, you're immune to all kinds of judgment. These are superpowers."

"I don't grow bigger by eating a red mushroom."

"Red *and* yellow," corrected his brother.

"You grow stronger by eating less poison," his father responded with a smile.

"Dad—"

"I'm serious. There are secrets that help you be better at life. It's up to you to figure them out, and to practice. Practice makes the master."

"Cactus?" came a voice from the floor.

"Use knowledge to open channels of communication. That's a good strategy. When you show people respect, they welcome you into their lives. That's power. When you help someone, they feel good about helping you. That's common sense, another power."

"Speaking of help," José said, sighing, "I'd like to play, too, please."

"You know how to win at this game?" Miguel asked his firstborn.

"Yeah. I kill my enemies and rescue the princess," Miguelito replied. "If they kill me first, I lose."

"There's one important thing you have to do to win this game," Miguel said, "and to win at life."

"Yeah?"

"First, you have to master the toy."

"The toy?"

"Toy?" Leo's head popped up from under the coffee table.

"What toy?" Miguelito asked.

"This little guy," said Miguel, pointing to the character on the screen. "Mario."

"Or Luigi!"

"They represent you in the game, right? The toy is you."

"Yes . . ."

"Find out what he can do, and all the power he has available to him, so you know how to overcome the obstacles." Miguel paused, watching the faces of his sons. "In life, you want to find out what you're capable of as a human, and practice becoming better and better at those things. You want to understand yourself, to see yourself as you are. Once you master the toy, you can master the game."

"This game?"

"And life. It's a game, and we're all playing it." Miguel smiled, happy to see them listening instead of arguing. "If we don't know the rules, we can't move to the next level. If we don't know our own power, we're at a disadvantage every time a challenge comes. We lose at every level, because we don't know the toy. We get defeated by ourselves, not the big monster."

"Bowser."

"Goomba!"

"Exactly. They'll crush us, only because we haven't taken the time to know ourselves."

The boys looked at the screen again, each coming to his own conclusions. At his father's urging, Miguelito cleared the program and set up a new game that included Mario and his brother, Luigi.

"Okay, *m'ijos*," said his dad. "Show me what you can do."

My sons are men now, and no longer squabbling over toys, but they've shown me what they can do in many ways. They are using the lessons of childhood to enhance wisdom. As they each adapt to a growing awareness, they are learning to be masters of their own story. If I never have the chance to talk to them again, I trust each of them to reflect upon the moments we shared, on my actions, and on things that I could explain only with words.

Of course, we're teaching our children all the time—even if we're not sure what the lesson is. Everything we've learned we pass on to our children, usually without stopping to consider whether it's true or what the consequences of our words might be. Children listen, even when they pretend not to. They listen, and they learn from us. Then they learn from their teachers, their friends, their idols; they also learn from the dream of the planet.

My parents did their best to understand something of life and pass it on to their children. I did my best for my own children, but I wasn't always there for them. I worked hard to make a living, and I worked hard on mastering this toy—this human. I created new strategies to discover more about the human mind, and my games grew into a talent for teaching. By teaching, I learned to use attention to its greatest advantage. Children see right away how important it is to win someone's attention. With attention, all channels of communication open up.

As a child, I had the attention of my parents, when they could give it. My brothers, however, had no particular interest in me. As the youngest, I hardly seemed worth their attention. Without many friends my own age, I had to amuse myself. In the pursuit of fun, inventing games to play by myself, I was surprised to discover that I could also win the attention of my older brothers—the people I most wanted to impress. If I was having fun playing a game, by myself or with a playmate, Jaime was usually the first to notice. He might sit down with us, to show us how to do it better. This would attract the attention of my brother Carlos, who would immediately try to compete with Jaime. Then Memín would notice and join in. As soon as my brothers had mastered a game, they no longer needed me to participate. Competing against each other, I was soon forgotten. So I would think of a new game, one with more challenging rules, and the same thing would happen again. Jaime and Carlos would get interested, then Memín. Sometimes my brother Léon got interested, too, and he would compete against the others. Without realizing it, I had stumbled on a way to get the treasured prize—the attention of my older brothers—and have fun in the process.

Humans are natural competitors. The first and most basic competition is our fight for attention. It starts early and never stops. I used to tell my boys about ancient Aztec games—games played in stadiums for crowds of thousands. These games were similar to our ball games now—two teams competing, usually by getting the ball through a hoop or over a goal line—but in those days the losers lost their heads, some say. Whether that's true or not, that idea got my sons' attention. The stakes were as high as they could be, I told them: life or death. The prize was survival, and the ball was the single means of achieving that survival. The same has always applied to the collective human dream, where the ball is represented by *attention*. It must be won and controlled, at any cost.

My students were taught the same lesson. As a shaman, I created many mythologies to assist my teachings, and invented different approaches to those teachings. I would say the same thing in different ways and in different contexts. I would then begin new traditions, with better and better results. As a parent, I used games to capture my children's attention—something that got harder as they grew older. By the time they were teenagers, they had very little interest in my point of view. They had many other opinions to distract them. The views of their friends became more influential, naturally, and they were fascinated by the lifestyles of celebrities, rock stars, and professional athletes. Like all kids, they turned away from this teacher and listened to others. They would come back, as I did when I returned to Sarita, but it would take some time.

Super Mario and his brother provided a great teaching tool, both for my sons and for my apprentices. I appreciate the merits of video games, because they explain life in visual terms. It's unlikely that most kids see gaming that way, but eventually the rules of any game are revealed as metaphors for life. In the process of growing and maturing, we discover our own strengths. We develop skills in school, at work, and on the playground. As we learn more about ourselves, we get better at life, moving from level to level in the human dream. I was in grade school once, and eventually graduated to high school. I was a medical student once. I graduated and got my degree. Then I practiced general medicine and surgery. With every graduation, I learned more about myself, making it possible to enter the next level (and a new world of people and challenges). Everyone's life is like a video game; every video game is a reflection of our reality and how we understand it. First, we must master the toy.

I was usually able to capture the attention of my children by meeting them in their own dream. Their interests became my interests. Their concerns became mine. They each had different interests and obsessions, of course, and their attention was constantly being pulled in those directions. My son Miguel has always been a sports enthusiast; soccer was his favorite game growing up. It was important that he knew his strengths when he played soccer, but also that he knew his role in helping the rest of the team. The *team* wins, not any one player. He may have wanted to be the goalie, but his strength lay elsewhere. To make the team stronger, he had to do what he was best at doing. Selfishness brings down the team and loses the game. Of course, we learn these things in many ways and through many different pursuits—even playing video games. Young Miguel was also intrigued by chess. He had watched me play with my brothers all his life. I don't know if I was ever as good as they were, but I reached a certain kind of mastery, winning local tournaments for fun. Playing chess with him, and many times with Dhara's son, I was able to teach important life lessons. Chess is a game of strategies, like every other. It encouraged the boys to imagine several moves ahead, and to predict reactions to each of their actions. It strengthened their own attention as they watched for mistakes and took advantage of the distracted attention of their opponents.

Humans live in a universe of consequences. Every action leads to a reaction, and some of those reactions penalize us. I had to teach this to all my sons. The importance of this lesson was evident to each of them by adolescence. As I explained to José, poor performance in school would lead to expulsion, which would go on his record. With expulsion on his record, he would be more vulnerable if he were ever accused of a crime. Time in jail would lead to unjust treatment by society, and then more complications, more injustice, and so on. Actions had reactions. It wasn't my intention to create fear, but rather to encourage my sons to see several moves ahead. The human dream is predictable, and most penalties are avoidable. José's favorite spectator sport was professional wrestling. He understood when I told him that it was all fake, a big performance to excite the audience, but that he could make it as real as he wanted. He could make it real, just so he could enjoy the game. Fun is the purpose of every relationship, every interest, and every interaction. Fun is the reward for our relationship with life, and we can avoid compromising the fun by seeing it all as a game, a dream.

My youngest son had his own interests. As Leo grew older, he became interested in poker. Gambling can victimize us if we allow it, and I wanted him to understand that it was a game to be enjoyed, like everything else; nothing more. In life, we assess the risks—some risks are worth taking, some are not; but we don't have to be afraid. We can practice using our attention to help ourselves. We can learn to recognize the fear in an opponent, and we can use our own confidence to change his or her strategy. If it looks like we're losing, we can wait for another hand. Patience and calm are strategies. Restraint is an excellent tool; but the best weapon against defeat is self-respect. Going against ourselves is the biggest temptation, a temptation based on the lies we believe. What do we believe about ourselves, and how important is it to protect the body we occupy? Where should the power of attention be, if not on saving this human, in any way we can? We may not have to face life-and-death competition, as did the game-players of ancient civilizations, but we can save ourselves from defeat by respecting ourselves—at any time, in any situation.

The dream of the planet is bigger than we are, obviously. It speaks from the mouths of billions of people, and its power is formidable. We are masters of our own dream, however. We are the strategists and the dreamers of our own reality. It's essential to understand the rules human societies set down, and to respect those rules as much as we respect the dreams of other individuals. The Toltec people honored the freedom of every individual to imagine reality his or her own way, and then to take inspired action. I have always wanted my apprentices to inspire themselves—and trust themselves to be masters of their dream and saviors to their human being. My best students gave me their complete attention and took action to change the way they dreamed. When they mastered one level, they moved on to the next. Actions created reactions. When they set out to become happier, they became happier. When they committed themselves to change, changes happened. When they were eager to learn as well as unlearn, they were able to master the toy.

Taking action resulted in the wonderful consequence of change. Where some people welcomed change, others fought against it. If the fun was gone for any apprentice, I encouraged the person to leave. We were born to follow pleasure. It was hard, at times, to let go of someone with great potential. It was hard to say goodbye to a friend, but I respected their choices. Freedom was always the reward—freedom from knowledge, freedom from themselves, and eventually freedom from Miguel. Whenever they wanted, they could ring the bell and leave. They could go home . . . or they could keep playing, conquer each new world, and win the game.

20

MIGUEL JR., SON OF THE SHAMAN, SAT BY HIS
father where he lay in cardiac care. He had arrived
in San Diego weeks before, having received an early-
morning telephone call saying that his dad had had a heart attack. Delivered initially to a hospital nearby, his father seemed
fine, was talking to family. By the time Miguel the younger got
into town, however, his dad was slipping out of consciousness
and unable to make sense of his surroundings. He was then
rushed by ambulance to a hospital in La Jolla, where some of
the best cardiac surgeons in the country were now fighting to
keep him alive.

With his father comatose and on life support, someone had
to take charge. Mike, as he was usually called, was the oldest
son, so he was the one the doctors consulted and advised; but
after nine weeks of coma, there appeared to be little chance
that Miguel Ruiz would survive. Any decision made in the

next forty-eight hours would determine the family's future. How could he be expected to take on this burden? At heart he was still little Miguelito, the boy who was good in school and obeyed his parents. His life was about the basic things—girls, games, and having fun. He was a kid. His father would have disagreed about that, of course. They disagreed about many things.

Mike felt his father's love always, but he also sensed his unspoken wishes. Miguel wanted his sons to continue his work and to benefit from his teachings and the wisdom of their ancestors. At twenty-six, Mike was still enjoying college. He was living in Oakland, away from his family, and away from anything related to the Toltec teachings. Having traveled with his father for years, witnessing firsthand the frenzy of power journeys, he was glad to be out of it. He'd had enough of shamanism. He'd had enough of men who wanted to be mystics and women who wanted to be wanted. He'd had enough of rituals and trances and spiritual excursions.

He loved the known world; he loved the dream of humanity. Knowledge excited him—the energetic exchange of ideas and opinions. His father had warned him about gossip and the spell of opinions, but what fun was it to have no opinions? Minds were made to meet, to share, and to conspire. His world was one of theories and memories. Knowledge was his craft, and college was an exuberant marketplace of ideas. He would stay in school forever if he could. In Oakland, people went to school, philosophized, dated, and drank. There, life was simple and the world was sane. There, he wasn't somebody's son; he was a guy with a girlfriend, a camera, and a healthy love for soccer and a six-pack.

Mike watched his dad's chest as it rose and fell to the flow of the respirator. He had never seen him weak, much less helpless. He had never really expected him to die, but it seemed that there was no escaping death now. Miguel's heart was too damaged, and his lungs were failing. His lifelong fear of drowning seemed a certain reality. Here, in this bed, connected to a dozen wires and monitors, his lungs were filling with water. "Take him off the respirators," some suggested. "It's time." If he said yes, how could he face his grandmother, who was determined that her youngest son must live? She was praying by day and holding sacred ceremonies by night. Her world was very different from his. Since he was a boy, he had watched her perform miracles, heal the sick, and conduct her rituals. Many times he had been her helper and translator, but what had he really learned from her? How could he assist her now?

His father would have shed some light on the situation for him. He would have asked for his full attention and made this a teaching moment. Mike smiled uncomfortably at the thought. His dad had been happy, even excited, at the prospect of dying, or so Mike had been told. In the first few hours after his heart attack, Miguel had had much to say, apparently—talking to his family, his friends, and a few close students. In those precious hours of consciousness, he had even signed the forms necessary to turn his houses and properties over to his sons. He had taken care of family business, all the while laughing, all the while sharing messages of wisdom and love. Now he was silent. Now it was up to his eldest son to act.

Mike needed wisdom now, and his father was no longer there to provide it. He needed advice from an elder, but Sarita had none to give. His *abuela* was in her own world. She was the warrior now, battling with the angel of death . . . but what did that mean? He had spent ages avoiding answers to questions like that—so how could he imagine what his grandmother was doing now, or who it was she fought in her nighttime dreams and daytime trances? She was in a different universe, but this real and present world wanted him to act. It demanded that he make a decision about his father's very existence. How would Sarita respond if he made the wrong one? What could he say to her then? What could he say to the teacher, the healer, the matriarch? What might he say to his dad right now?

"Say whatever you want," Miguel urged. He was sitting up in bed. Gone was the ventilator tube. He was free of sensors and wires, free of life support, and smiling happily. He was a vision in his own mind, if nobody else's, wearing his hospital gown and a Padres baseball cap. They had gone to a hundred games together, he and Mike. How great those Sunday afternoons were! The memory of those times just might encourage his son to talk to him now, and to trust. "Go ahead," he offered brightly. "I'm listening."

The young man seemed deep in thought—face frowning, eyes down, and one leg bouncing nervously. Yes, Miguel recalled . . . thinking was torture. He'd tried it once, after years of doing without, and vowed never to repeat that mistake. Thinking was agony. Miguel shook his head and reached out to feel his own bare toes. The gesture required no thought. Thinking was nonsensical, at best. Of course, it all made sense to the thinker, just as another round of tequilas made sense to the drunk. After years of ingesting poison, a man comes to believe it's medicine, or the friend he can count on, and he refuses to live a day without it. As it turns out, anyone can live without poison. Anyone can live without thought, without the persistent drone of words and commentary. People can live without the noise in their heads and its emotional residue. He had explained this many times to his sons, but they would have to wage their own private wars against the noise. They would have to fight their own battles with knowledge, and win. They would have to listen, finally, and not believe.

Miguel observed his child, his firstborn, and smiled wistfully. It was hard to see his son so tormented, but he felt no judgment. There was only love, spilling from him like a river flowing into the sea. He had loved this boy since his birth, since he picked him up and lifted him to the sky, proudly proclaiming him to the universe. He had loved him through all the changes. He had loved him through all their disagreements, confrontations, and separations. He and his son had spent too much time apart, but they were together now. They had sometimes failed in their communication, but they would not now. What Miguel had said to him, often and sincerely, his son had heard. Perhaps the boy had resisted, but he had *heard*. There was wisdom in him now, and it would serve him in these next difficult days and all his remaining days. His father's love was beyond question. The love that life provided was impossible to avoid. As father and son sat together, machines humming nearby and nurses chatting softly in the background, love washed away divisions and cleared memory's wreckage. He saw his son relax. He saw him catch his breath, exhale, and quiet his mind. After a few minutes, Mike lifted his gaze, and his eyes met his father's. Miguel waited, wondering what the younger man could see.

"Talk to me, *m'ijo*," he eventually said to his son, but there was no response.

They were too far from each other, he concluded. Even with the best of intentions, even with the baseball cap on his head, his presence would be too subtle, too remote . . . so he let love bridge the distance. He drank in the wondrous sight of his son and was gratified. He felt the young man's mood, his worry, his fear; and he calmed a mind clamoring for answers. This was something he had done many times as a shaman. He could wipe all the painful thinking away and empty a student's mind. The reaction was disorientation at first, and then surrender. Sometimes the body collapsed, like an animal finally released from the whip. These moments didn't last, of course.

They came swiftly and were swiftly gone, leaving the student with a taste of the freedom he or she might someday achieve without help.

He wanted his son to act freely, and to trust himself completely. There was no need for worry. There was no need for thought. Any decision was a good decision; any action was life's action. Faith in himself was faith in life, no matter the outcome. Conversations between them might never happen again, but the memory of words spoken and things shared would remain alive in Mike always. He would remember the fun, the family outings, the ball games. He would remember the lessons of the Super Mario brothers . . . mastering the toy in order to win at the game.

"It's time, *hijo*," Miguel said. "You know what this toy can do. Now it's time to master life."

He smiled at the boy—a man now, making decisions for his family and preparing to put the secret wisdom into action. Magic was the natural motion of life, and father and son were making magic in this moment.

"I love you," he said to his oldest son. "Listen now, and play with me again."

Don Leonardo was sitting in the front tier of bleachers at the Army and Navy Academy north of San Diego, where Leo, his namesake, was playing football with his classmates. The boy, not yet graduated from high school, was reeling from the possibility that he might soon lose a father. This one was born after his own demise, of course, so there was only a slim argument to be made that a bond might exist between them. Still, as he watched the boy running determinedly through the mud, dodging opponents and careening after the ball, he saw a familiar spirit. Miguel had taught him the lesson of attention, he could see. Today, the boy was applying his attention well. The human dream was like this: a game of win and lose and win again. The ball represented the attention, which humans fought for above all else.

There was a time in this boy's life when he didn't receive the benefits of his family's attention. By the time he was a teenager, still living with his mother in Tijuana, Maria was finding him difficult to control. Miguel had stepped in then, taking him to the house he shared with Sarita, and to a life in California. This meant a new environment and a change of culture. This meant perfecting new language skills and learning the rules in his father's house. He was sent to school—to this school, which was close by but offered him a chance to flourish on his own. From what the old man could observe, it had worked. The boy had friends, he had responsibilities, and he was free from the criticisms of his family. Leonardo was well aware that there were no judgments made against the boy, but it was impossible for him to know this. Children imagine that all eyes are focused on them, and that they are the target of general disapproval. Sadly, it was true that most people gossiped and judged—but how to tell a young boy that there were no judgments as unforgiving as the ones he cast against himself?

The boy Leo was sitting on the bench now, resting, while the game continued without him. He had a towel in his hand and was rubbing his head absently. His hair was an improbable shade of red, like Maria's, and it distinguished him in an intriguing way. Rojo, his friends called him, and it was plain to see he liked the distinction. He liked having his own dream; he enjoyed living in his own world. There would come a time, the old man mused, when he would beg to be set free—to make his own mistakes and to fail or succeed without his father's advice or protection; but the time for autonomy had not yet arrived. Today, amid the yells of friends and the solace of a warm rain, life without his dad was just an idea, an anxious thought, a cloud content to linger on the horizon.

His father was in critical condition and his own future was in doubt, he knew; but his family was there. His grandmother, always strong and always comforting, was there. Mike was there, making the choices that needed to be made. His brother José was there, staying away from the gossip and the noise. José was married now, and the house he shared with his wife was Leo's second home. It was a safe haven from relatives who asked too many questions of him—questions that had no answers.

Young men never had answers, don Leonardo noted. They asked few questions, not wishing to seem ignorant, and they had no answers. They boasted, they demonstrated contempt for their elders, and yet they knew nothing. The old man preferred to think that he himself had been a different kind of youth, but that was likely untrue. In his own adolescence, he had run from his family, thinking he could create another, better one. He ran blindly, making a thousand stupid assumptions as he went. Life had intercepted at every turn, repeatedly saving him from self-destruction and from jail. Life had saved him from war and from the hungry jaws of death. Life had given him music, women, love. Life had given him children, and children had given him a better understanding of life.

Don Leonardo, the echo of an old story, looked across the field at the youngest son of don Miguel and was comforted. This boy, the one who carried his name, would live through the challenges ahead, because life said so. He would stumble, thrive, and ultimately survive, with the winds of creation at his back. Father or no father, he was blessed. He had been touched by good fortune since he took his first breath—although he might claim otherwise. Faced with loss, he would grieve, of course. He might curse the heavens and cry himself to sleep at night, but it was clear, even to this dead man, that the boy was blessed.

Leo was back in the game now, his red hair glistening in the rain. His feet kicked up mud, spraying him from ankles to ears, as he called to his teammates for a shot at the ball. That's right, the old man thought, nodding. Attention is the prize. Where is your attention now? Where will it be, should your life suddenly change? Will it be on bleak injustice, or will it be on gratitude? Yes, gratitude—for the many riches of your existence, for love that is given freely, abundantly, and without terms. How you respond to life's abrupt changes will make every difference, and will eventually turn a wayward child into a righteous man.

Don Leonardo stood up and brushed raindrops from his creamy suit, glad to have been a part of this moment. There was more to a man's life than memories, certainly. There was present possibility, along with the slow, percolating process that brought it to life. There was love, whose seeds required only a moment to plant and a lifetime to grow. It was up to each man, free from the constraints of childhood, to nurture that love in himself, making its harvest evident to the world.

Don Leonardo, the wayward son of his wayward father, stepped off the bleachers and strolled away from the field, heading toward the yellow glow of evening and back to the spaces of memory his daughter now occupied.

◉ ◉ ◉

Mother Sarita sat comfortably on a fallen tree. She admired the lacy veil of fog that draped the hills, happy to feel the sun's warmth again and relieved to feel the certainty of the ground beneath her. Her excursions had weakened her, endangering her physical health. She had tried too hard, driven by her own eccentric ambitions. She had taken reason to its farthest limits, frightening her loved ones and compromising her objective in the process. Some rest was needed, she admitted, and she would get it. Her body was safely tucked in her own bed again, and she was dreaming restful dreams. As if to make sure of that, she looked around, inhaled the cool, moist air, and congratulated herself on conjuring such a serene landscape.

In the midst of her soaring visions much earlier that day, Jaime had dropped by the house and found her still seated in the big chair. He woke her gently and prepared a hot bath before leaving again. More relatives arrived in the afternoon but were gone before dinner, allowing her to get to bed early. It was time to see things in a clearer light. It was time to evaluate, and to make better choices. This errand must not fail.

She wasn't sure why she was here now, in a canyon embraced by cliffs that held a hundred primordial caves, but she was glad to feel like herself again. The place looked familiar; she recalled taking day trips to natural settings like this when she was young, dreaming of magic and enchantment where only animals lived. She had taken many long walks back then. She had always enjoyed the outdoors; she had loved the sun, the meadows, and the smell of wild herbs and summer blossoms. She had liked digging for healing roots and dipping her toes in cold, bubbling creeks. She had liked picnics with her sisters, eating the bread and cheese that they carried with them in worn leather satchels. She had liked lying in the soft summer grass, counting butterflies and chatting about boys.

"Sweet gods," moaned La Diosa. "Are you twelve years old again? Have you traded your wings for a petticoat, old bird?"

Sarita cringed. She was not surprised to hear that voice again, only disappointed to have this peaceful moment interrupted. Sighing in resignation, she looked up at her visitor. She had begun to accept Lala for what she was. Sarita recalled her most recent conversation with Miguel—had he been real, had those words been his?—and knew that she should dream this differently. She must be more aware of what was, and not so sensitive to the scenery. She could feel the pull of this woman, as strong a pull as the memory of her son, and felt his will working through her. He was here in everything. He lived in the heat of the sun and in the strength of the hills. He was watching, waiting for the ultimate shift. *This errand must not fail.*

"These are private reveries, my dear," Sarita responded politely. "This has nothing at all to do with you . . . nothing to do with our arrangement."

"Odd," Lala replied. "You vanish, sailing heedlessly into delusion—"

"No delusions."

"—and then you return, wings scorched, with a beak full of childish lies."

"What lies?"

"Did you not just say to me, 'This has nothing to do with you,' or was that just an old crow clucking?"

"This has *nothing*—"

"No? Then why is the shaman here?"

Lala pointed to some hikers who were making their way along a deer trail just below the cliffs. Sarita stared at the group for a full minute before recognizing her own son, a young man in peak physical condition, leading two small boys through the wilderness. Their approach was suddenly obscured by a stand of cottonwoods, and then the three of them appeared again in full view, with Miguelito taking the lead.

"Madre Grande," Sarita uttered in surprise.

"Who?" asked Lala.

"Madre Grande," Sarita repeated. "In California. Near a

monastery, I believe. He took Miguelito and José to this place when they were small—just the men of the family. It was a shamanic journey for them, and such a wonderful memory!"

So, the memories were still beckoning. All was not lost. She put a finger to her lips, silencing Lala, and listened. She could hear the oldest boy, Miguelito, singing at the hillsides, making echoes that chased each other around every rock and cloud-draped promontory before they circled back to where he was, filling him with delight. He might have been eleven years old, but she wasn't sure. Between bursts of song, he chatted to his father excitedly, while José followed in a kind of dreamy silence.

"Look at my superpower, Pops! I can make the mountains speak!" With that, Miguelito warbled a song and waited for it to circle through the hills again. "Is it power, really, or is it magic?" he asked his father. "Because I'm also a magician."

"Yes, you are," agreed Miguel. He was walking at a slow pace, taking in the feel of the place, his hands held together behind his back. He always walked that way; and behind him, José was doing the same. "Power is potential energy," he said. "What people call *magic* is power in action."

"Action!" his son shouted, and the hills shouted back, "Action!"

This somehow proved the point, which gladdened the boy immensely. Turning to smile at the others, he saw a red-tailed hawk sweep by, saw his younger brother leave the trail in pursuit.

"Pop!" Miguelito called, "José is gone; he's in the bushes!"

His father had stopped at a divide in the trail, his imagination following the course that each path might take. "Here . . . this is the path that leads up to the rocks," he concluded. "Where is your brother going?"

"Like he'd tell us!" the older boy said, snorting. "When does he ever say anything—to anybody?"

"Then let's keep walking. He'll catch up." He turned up the new trail, and his son raced ahead, pushing him aside. "It's good that you brought up the subject of power," Miguel started, but Miguelito was shouting to him again.

"Pops! Look! José found it! The cave!"

José had scrambled up the hillside toward a wall of boulders and into the low-lying mist. As they watched, they could just make out his little-boy shape slipping through a smooth crevice and disappearing into the cave. Miguel and his oldest son ran up the twisting path to find him.

"Here!" cried Miguelito, who reached the spot before his dad. "He's in here!"

Just then José came out, flushed with excitement, his eyes wide with wonder.

"I followed the *halcón*," he panted, "and it brought me here."

"Thank you, *m'ijo*," his father said. Smiling, he contemplated the flat, wide surface of a boulder beside them. "Shall we do the ceremony out here, where the sun can find us?"

The boys nodded enthusiastically. They turned to admire the narrow valley below, and all three fell into silence, watching, waiting. The hawk whistled past again. Then, in a magnificent burst of light, the sun shot through the mist and struck them. The two boys stood transfixed as their father began to speak.

"Life, the supreme artist, takes an action," he said, "and the result is magic. To know yourself as life is to know yourself as the magician . . . the artist . . . the Toltec."

The shaman raised his arms in the air, bringing his palms together over his head. He looked toward the boulder, and his sons turned to see his shadow outlined against the flat rock. As they watched in amazement, Miguel began to move his entire body in snakelike fashion, causing the shadow to come alive.

A huge rattlesnake seemed to rise up on the rock, writhing, taunting. The boys swallowed nervously, then began to move slowly in imitation, their eyes focused on the monstrous snake. As they moved, two more snakes rose up to dance, and from the surrounding hills came the sound of rattles, the music of a thousand gourds. The strange ritual continued; and as Miguel held the trance, his shadow suddenly sprang to life. The snake leaped, and the boys startled, jumping back in fear. Darting nervous glances at each other, they found their courage: they put their hands together above their heads as Miguel had done and moved in rhythm with their father once more, staying as close to him as they could. The spell strengthened, wheeling the world into a deep and deliberate trance. The valley hissed and rattled . . . louder, louder.

"Do you hear that?" Miguel asked. "The mountain is welcoming us."

Nodding somberly, the boys kept a close eye on their shadows. The snakes coiled and twisted, absorbed in the motion of life. Mesmerized by his own shadow, little José soon began to dance to unheard rhythms, closing his eyes and sensing the magic. He hissed softly to himself, taunting his own fears, as the sky grew brighter, engulfing him in a burning, blissful white light.

José Luis, who was named after his father's father, was sitting on the beach. It was a gusty spring morning in Malibu, and the Pacific Ocean growled restlessly at his feet, but he was dreaming about a spiritual desert. In sacred storytelling, the desert has always offered truth and revelation. Away from society, unhampered by its rules and exiled from its solace, a warrior finds the isolation he needs. With no visible signs of life to re-assure him, he confronts his biggest demon; in the desert, alone and comfortless, the spiritual warrior faces himself.

José, the second of three sons and now an adult, wasn't sure when his father had first told him about this desert of the mind, but he was prepared to go there now. He had spent his young life feeling isolated from friends and family. He had always felt out of place in the world-dream. His high school years had taken him to the verge of annihilation—drinking, taking drugs, and separating himself from those who could offer help. He had hardly spoken during that time. He had hardly responded to the polite inquiries of friends. He had seen nothing in people that moved or motivated him, and he had seen nothing redeemable in himself. He'd been a lost boy, an orphan within a big family, and his pain had been impossible to conceal. He had slipped so far away from his father that even now it was difficult to imagine Miguel as he was. He could remember his wisdom, however. His father's words had been at war with José's self-rebuking thoughts every minute of every day. The war had still been raging when he was twenty, and his own voices had been winning. The noise had been defeating him—until Egypt.

José and Judy had met on his father's last power journey to Egypt, and they'd fallen in love. Marriage had saved him, but the pressures of a relationship had brought out old fears, the fears of a dejected adolescent. Euphoria and anger still played against each other, and the result was considerable drama in a dream that should have felt safe. Then, in the dim light of dawn one winter day, the news came—his father had suffered a heart attack. Don Miguel was preparing for his death, and he wanted his children by his side.

José had arrived at his father's hospital room in tears,

begging him not to leave him at this difficult time. He cried openly, his tender heart already breaking and fear commanding his words. His father's expression stopped him. Miguel had been so happy to see him, but now his face became stern, as if the two of them were standing in the principal's office again. Miguel had never been the type of father to scold. A look was enough. A long, hard look could make José feel like a child again, guilty and repentant all at once. There, in the emergency room of the hospital, his father gave him just that look, and it silenced him.

"Is this the way you celebrate the death of your father?" Miguel asked. "Go! Leave the room! Fix yourself! When you're finished, come back, because I have important things to tell you!"

José, sniffing back the tears, did as his father had asked. He left the building, searching for solitude. He found a spot by a small tree that stood bare in the winter chill, and was glad for its meager company. Replaying the conversation with his father in his head, he recognized his own selfishness. "Don't leave me!" he had cried. "Don't die, Pops! I'm not ready!" He had pitied himself, fearing for his own safety and obsessed by his own needs. How quickly he had buried his father and submitted to grief—even as the man sat there, alive and happy to see him! Was this all he had to give the man who'd granted him life, love, and the full power of his faith? Was this how he repaid his teacher? Was this truly the way he would celebrate the life of his father? As his tears dried, resolve began to strengthen him. He left the little tree and returned to his father's bedside.

"Pops," he stated evenly. "I was being selfish. I was afraid for myself, and not even thinking of you." Sitting on the bed, he took his father's hands in his. "I can do as you say—I can rise above fear. I can hear knowledge speaking, and choose to disbelieve it. I can hear the way people spread poison, and choose to ignore it. I'm ready." José looked directly into his teacher's eyes and said, "I'm with you now."

He remembered how his dad had beamed then, saying that he'd intended to talk about those exact things. The lesson was over. They sat together, as Miguel shared his vision for his sons and spoke of his enduring love for each of them.

That conversation seemed long ago now. Miguel had been in a coma for two months, and the doctors now conceded that there was little hope. José had tried to remain composed, ignoring his own troubled thoughts and the gossip of family members. He avoided the hospital these days, choosing instead to spend time at home, to be happy, and to look after his younger brother, who was gripped with worry and unable to speak about it.

José recalled the years when he couldn't speak either—when there had seemed little sense in talking. So much had happened since he was a sullen teenager, trapped in his own private hell. Falling in love had changed him, giving him confidence and a surprising curiosity about the future. He had discovered an appreciation for life. His imagination was on fire. For the first time, he had a strong desire to learn. In these last weeks he'd started listening to audiotapes of his father's private classes and public lectures. He'd begun to hold his own seminars . . . in the bathtub, and to no one but himself. He liked to hear his own voice booming against the tile walls and resonating far into future dreams. The boy who'd once had no voice, who'd spoken to no one, was learning to speak at last. He was learning to heal his angry heart and to lift the hearts of others.

José could see himself as he'd been—a parasite, taking his time to devour the human he occupied. He'd committed crimes against himself. He'd cowered in the dark and believed in his own lies; but he was starting to develop a taste for truth. He was beginning to love life, and himself. True to his promise, he was with his father now.

His father, for his part, was with him, too. The days and weeks of anxious waiting had provided uplifting moments along with concern. José felt almost joyful sometimes, as if he and his dad were having a good time at the county fair, on a roller coaster and sharing cotton candy. He couldn't explain it, and his father would have laughed at him for trying. "Why ask why?" he would have said. "Just have fun!" Their communion, however it happened, filled him with love—and, bit by bit, a deep understanding. José had learned to recognize the voice of knowledge and to challenge its deceptions; but he was still tempted to doubt and to descend into darkness. It was time to concede the struggle and acknowledge the pain it was causing. It was time to face himself again.

Ocean waves slammed upon the shore like rowdy children begging for attention. The wide sea rumbled into the distance, and into timelessness. José took a deep breath, let the memories subside, and opened his vision to all points of view. It was possible, he knew, to see as life sees, and be free.

Practice would make the master.

21

EMMA HAD FOUND HER EQUILIBRIUM AND WAS feeling better. An hour or so in the mirror room usually did that. She was exhausted from days and weeks of exultant upheaval. Exultant—now there was a mystery! From the time of Miguel's heart attack until now, so many weeks later, she'd felt only joy. The man she loved was in a coma, unlikely to survive the crisis, and yet she was at peace. She missed his voice, his touch—and yet she didn't miss him at all. He was more present now than he had been in the eight years they'd known each other. He was laughing with her and at her and poking fun at tragedy. His words spun around in her head—teaching, chastising, comforting. He was in the passenger seat as she drove back and forth to the hospital. He lay close to her at night, and she woke up to his smile every morning. This persistent joy chased away the doubt. She'd never felt better or more content—and at a time when everything seemed to be going wrong. She'd never felt stronger in her trust or more in tune with life.

Maybe this was what freedom felt like. "You don't have to be you," he used to say. How many times had he said that—something so simple, and yet so difficult to grasp? Miguel had no obligation to be anything to anyone now. He might soon be free from matter itself. He might soon be home.

Thinking of her own dream, so precarious without him, she wondered why all of this was happening. And why—why?—was she feeling such a sense of calm?

"Why ask why?" Miguel answered in the sweet voice she knew so well. He wasn't with her in the little mirror room . . . but, as usual, he was with her.

"I knew you'd say that," she responded.

"You know things."

"I still want to know things," she agreed. "I'm hopeless."

"I hope you're hopeless," he said. He smiled, and she smiled with him. He liked to call hope the biggest demon in hell. It enticed and beguiled, but delivered nothing. Hope wasn't making its usual mischief, thankfully. She had no hope for any particular outcome, nor did she expect things to go any particular way. It always came back to surrender.

"Do you like my gift to you, sweetheart?" Miguel asked.

"Your gift? You mean this joy? This cluelessness?"

"I mean Miguel's legacy."

"Legacy," she mumbled, frowning. "Legacies belong to the dead."

"Legacies belong to the living. When someone stops existing, what is left of him?" he urged. "What remains?"

"Things he said today," she said, echoing the refrain of a Beatles song.

"Exactly," he said, chuckling. "Things he said."

He and Emma had always had the same music playing in

their heads; this had been true since they first met. Music had sparked their romance—music and life's chemistry. He remembered many sleepless nights, singing to each other. Sometimes they turned it into a game. One would start a song, then stop in the middle of it. The last word they sang would make the first word of the next song, on and on, until they could sing no more, falling asleep to the lyrics of a hundred classic melodies.

"My legacy to all of you," he said, "is the memories you have of me—each different, all of them dreams of your own making. My legacy is the teaching, however it is interpreted. My legacy to everyone is like a music library, custom-made for the listener."

"My memories of you are all musical."

"Really?"

"Okay, and physical," she said, yearning to touch him. "And emotional." It would be dishonest not to include the searing heartbreaks she had experienced since she first fell in love with him. "Occasionally painful," she added.

"You've used me as an excuse to hurt yourself," he commented. "This can stop now, my love."

"It will stop for sure," she offered, "if you come back."

"Oh, no," he said, laughing. "You sound like Sarita!"

"I'm not Sarita, though," she said. "I'm . . . I'm your . . ."

Sighing, she gave up. Who she was hardly mattered now. She was untethered, set adrift on a sea of mystery, and she had no answers. She felt his love more than ever now, and wondered how she could have turned such unequivocal love against herself. What kind of lies had she believed all these years, to make love seem so dangerous? Love with conditions was the opposite of love, he'd said. It was the warped reflection. Now was the time to end the distortions, while he was still near, lifting her with him into paradise.

She gazed at the mirror in front of her, seeing a woman who looked vaguely familiar, but who bore no resemblance to the woman he had found—and resuscitated—so many years ago. She saw a woman without a story and without fear. Most significantly, she saw a woman who was intensely happy and without hope.

"This is intolerable," Gandara whispered self-consciously, trying to squeeze his shoulders into the mirror room where Emma sat in silence. "What sort of magic is this devise meant to conjure?"

"Not magic, I presume," said Eziquio, wriggling within the small space, "but mood." He had left his sombrero and boots outside the mirrored door, but he was still uncomfortable and struggled to find a suitable position. "I believe she is conjuring a mood."

"Tell her that my mood is degrading rapidly," Gandara groaned. "Shall we alter the landscape, *patrón*?"

"We found her here," Eziquio reminded him. "There must be a reason."

"Ah, now reason has joined us! One too many characters, if you ask me."

"Be patient, my friend," Eziquio said, peering over his bony knees to survey the small space. "This may be interesting."

"This is intolerable!"

"Look around you!"

The two men ceased their floundering to consider where they were. They had crawled into a handcrafted room of eight mirror panels, all framed in polished oak and hinged together with an artisan's skill. It was a piece of furniture, it could be said, but one that had been built to accommodate one person . . . a peculiar someone who might enjoy contemplating the world as life's magnificent reflection. At this present moment, all that could be seen within the tiny mirror room were reflections of the woman apprentice. She was the only living thing here, but her reflections dominated the space. One mirror image enveloped another, then another, and on and on, spiraling into infinity.

Gandara squirmed in his confinement, but he had to admit that this space was ingenious. He wondered why he had omitted this from his education—allowing simple mirrors to create an infinite universe. Placing oneself in the middle, one could see the real world as a mere projection of mind. One could imagine life, infinite and mysterious, visible only in the reflective surfaces of matter. This was good. This was disturbing and wonderful. If he and the others had taken more opportunities to dream like this, they might have been formidable even in childhood. He smiled at the notion, rearranging his beefy shoulders within the confines of the octagonal box.

He noticed that Eziquio had finally settled on his side, pressing his backside against the glass panels, but finding no appropriate deployment for his spindly legs. Gandara, with most of his corpulent body outside the room, had managed to push only the top half of himself into the cramped space, and was presently resting his round, unshaven face in both hands. Though displeased, he took the advice of his friend and looked carefully around, as Eziquio was doing. Wherever they looked, they saw the same thing. Emma, seated quietly with legs crossed, was surrounded by countless copies of herself, all of them sitting exactly as she was. The two men craned their necks this way and that, trying to catch a view of themselves around the many Emmas, only to come face to face with each other. They both turned away, disgusted.

"As I said," Gandara repeated pointedly. "Intolerable."

"Perhaps you prefer the tombs of Egyptian pharaohs?"

"Once was enough, *gracias*. Still, the chambers provided a better view."

"*Hombre*," said Eziquio, "the view here is spectacular. See beyond this human, beyond these reflections, to the infinite dream."

"Take an adventure, you say?"

"As in the days of our best ruses, my friend."

"Life was a drama then, was it not? An opera."

"Complete with divas and buffoons."

"But we held the baton, no?"

"We held the baton, yes."

"Ah," Gandara said, remembering how he used to see the dream of humanity as an orchestra, a vast composition of sounds and tempos. How he loved to guide the music, how deftly he held the baton! His pudgy fingers floated to the rhythm of unheard melodies as he recalled those good times. He was silent, closing his weary eyes and letting his head rest on the soft carpet beneath him.

He loved the feel of existence, and he missed the power to see past its imaginary walls. He could sense invisible things back then, when he was alive in his body. He fancied he could hear people thinking; he knew their intentions before they did and could easily anticipate their actions. As a mother knows her infant, he knew the language of the human mind. As a man knows his lover, he sensed life's breath on his flesh, and he responded to every lustful yearning. He could see life's currents ebbing and flowing from everything to everything. Why, he could see life shooting from his own fingertips! Ah! What a joy it had been to be alive!

"Take a closer look at the mirrors, Gandara," Eziquio said. "Look up, my friend, and see how each mirror holds an array of memories, and how every memory is an opus, yielding to the *obra maestra* that is one single human life."

The fat man lifted his head and inspected the glass panel nearest him. Thankfully, he thought to himself, there was no sign of a scruffy old Mexican man with red eyes and bad breath. There was only the woman, sitting serenely, motionlessly, beside him. He could see many, many versions of her. He could sense the pattern of her memories. There were innumerable light projections from other mirrors, bouncing her image around this little room of infinite perceptions and telling a million stories—stories that warned him not to believe just one. As the enchanted moments ticked on, every story-picture came to life and moved to a kind of music, becoming a dance, a play . . . an opera.

<hr />

The mirrors . . .
    When I first added mirrors to my teaching, I didn't want Emma to participate in that kind of meditation. She was too avid a dreamer as it was, I felt. Remembering my own experiences, I thought she might go too far, enjoying her time with mirrors so much that she'd lose interest in reality. She did as I had done, however, and spent long hours there. Sharing my dreaming moments with her in this way, I see that she uses the mirror room well. She dreams with me. She dreams in creative ways, and when she has seen enough, she steps away. The beautiful little room of wood and glass is serving her especially well now, when revelations come swiftly and steadily. It consoles her during these days without a teacher, at a time when her future, like my family's, is so unclear.

    Don Eziquio is also serving her well. Old tricksters, it seems, can be the best allies in our quest for awareness. The more they argue and posture, and the louder their inanities become, the more we're able to understand what we're doing to ourselves. Hearing the mind's incessant chatter brings attention to the dangers of blindly believing. All the characters in our lives are imagined, just as our inner voices are. The real people they represent are nothing like the impressions we have of them. This is obvious to someone who sits quietly in a mirror room, where all reflections seem distinct and familiar but are not us, are not real.

The year 2000 was not the catastrophic event that some had predicted, but the changes that occurred in my own world were significant—a few people would even say catastrophic. The power journey to Egypt had left me uneasy and restless. I saw clearly that I had changed, that my investment in my present dream had lessened. To excite and renew my interest, I needed to find other ways to create. My first book had been published, and it included much of the wisdom that my apprentices were familiar with—but practice, as I always told them, made the master. I had often said that I would be their last crutch, the last psychological support they would have to abandon before they could fly on their own. Like Dumbo's little feather, crutches help people believe in themselves, and help them survive dynamic changes. Transforming our belief structure is the most important kind of change we can make, and in many ways the scariest, so a little help is good. Putting faith in the teacher helps. With every little change comes a big reaction, so a new mythology helps. White lies and harmless justifications help, too—until it is time to fly.

My students were lying to themselves less than they once had, but it seemed impossible for them to stop gossiping about themselves and each other. Their struggles with self-importance continued. Some needed to feel particularly important, so I had given them titles. For many, I used descriptive nicknames, encouraging them to imagine themselves differently. Like all crutches, these were meant to be discarded when the time was right—that is, when wisdom and awareness made distinctions unnecessary. The Four Agreements also helped their progress.

Seeming so simple at first introduction—be impeccable with the word, don't take anything personally, don't make assumptions, always do your best—these agreements made huge shifts in awareness. With every application came new revelations, and with every revelation came more awareness.

As wonderful as my interactions were with my apprentices, I was disappointed in the results of my efforts over the years. I saw that many of them were simply trading new superstitions for old. Jealousies remained strong, and selfishness was still evident. Generosity seemed difficult for many to give or to receive. We cannot give what we do not have, and for many of my apprentices, unconditional love was unfamiliar, untried. Many imagined themselves to be great teachers, a goal I encouraged, but their pride made that goal seem unrealistic. So, after Egypt, I made changes.

I started by doing away with titles. No one had an advantage over anyone else, even if that advantage was only in their imagination. There would be no gossiping. There would be no selfish actions. Each one of them was a potential warrior, capable of facing the battle within himself, but they'd all devoted too much energy to fighting outward battles instead. They would have to confront their own lies. "Don't believe me," I would remind them. "Don't believe yourself . . . and don't believe anyone else." Not believing their strongest opinions was the most powerful tool for awareness. Not believing their own thoughts, and the stories they'd created, was the best path to freedom. They fed on superstition, and my message to them was one of common sense. Appetites would have to change.

There were many reactions to this. There was excitement, but there were also hurt feelings and resentments. Seeing this, I soon realized that I would have to go further. I would have to tell them to go, and to use the tools I'd given them to create happier lives. In too many instances, fanaticism was replacing the desire to learn and to change. As I had done many times as a child, I challenged myself to invent another kind of game, this one with better results. First, I would need to end this one. Just as I had done years before, leaving medicine to investigate new ways of healing the mind, I would have to put shamanism aside. I would have to renounce power.

Few would ever ask what I meant by that. Few people truly want to understand power, preferring to accept common assumptions. And yet it's important to see the true nature of power. I renounced power because of the effect the very idea of power had on my students, and the zealotry that my presence incited. I had made my personal power evident to my apprentices. They had seen how I shifted perception, soothed fears, and healed sick bodies. They had seen that the force of life that flows through all of us can be used to enhance our relationship with life. We each have access to total power and the ability to change our reality. It was essential that my apprentices learned to do these things *for themselves,* with no expectations. It was important that they knew themselves as the saviors that they were.

We are life. We are the result of life's power, and we are the channels through which that power courses. We want truth, but we reach for more knowledge instead—and then we must defend what we believe. Knowledge makes a small impression on the world compared to the power of truth—even when it serves life, elevating awareness and creating dreams of impeccability. How can knowledge serve better? How can we stop it from doing harm and creating conflict?

First, we can see knowledge for what it is—all the agreements we make about reality—and get some perspective. Then we can listen to ourselves. We can modify both our thinking and our emotional attachment to thought. We can win the war in our heads, one that has gone on long enough. We can nurture a belief in ourselves and drift away from the crowd.

It could be said that, after Egypt, I drifted. During this time of transition, I had a chance to work with José and oversee his spiritual education. I wanted to put my attentions there. I wished for a new way to play with life. I experienced the peace of nonexistence in Egypt, but I had to begin again in this life, this present existence. It was important to see the next step, and to keep myself walking steadily through the human dream. Coming back from that expansive peace, I saw humans more clearly. I saw them as exceptional creatures under the tyranny of knowledge, helpless to change their circumstances. The human animal is helpless, that is, until the mind decides to change and until knowledge has lost its supreme authority. I was aware that I wasn't Miguel, that Miguel was just how I described myself to others; but whatever I was, I was in a body that needed my presence and my guidance. There are a few simple things the body must have to live. The mind, on the other hand, needs nothing. It invents needs, imagining that it's made of actual matter. It is, instead, dreamed by matter. An exceptionally aware mind understands that it serves the body's needs and the body's communion with life. A mind aware of itself is willing to listen to its own voice, and to change that dialogue for the sake of the human being.

If I were ever to teach again, every student's discipline would need to include direct conversations with the main character in their story. They would need to look in the mirror and an-nounce themselves. They would need to separate body from mind in their perception—in other words, to recognize the difference between knowledge and the very real human it oc-cupied. This was the only way *self*-awareness could have true meaning.

Months after the journey to Egypt, I began to feel incredibly free—as free as I had ever been. I was in love with life once again and dared to imagine that my love would inspire a vibrant new dream. So I looked ahead, trusting life's generosity as I always had. I was alive. I was eager to play with intent again. I was back, and ready to begin a different kind of game.

22

M Y NAME IS TOM. I AM A PREDATOR. MY FAVOR-
ite food is anger."

The classroom was walled with mirrors. All
eighty students, lined up side by side, stood in front of a mirror
as they spoke to their own reflections. They had been told to
get out of their chairs, move to the mirrors, and address their
reflections. It was harder to do than they had expected.

"My name is Lynda," said a tall woman, looking resentful.
"I am a predator, and I have a taste for . . . injustice, I think."

"My name is Veronica," another one said. "I am a predator. I
hunt for . . . I mean, I like to eat . . . shame."

"My name is Tony," said the man standing next to her, leaning close to the glass for a private talk. "I am a predator, and
I have a really big appetite for self-importance . . . and judgment . . . and pride, of course."

"My name is Ann," said another more comfortably. "I am a predator, and I can't resist jealousy. I eat it all the time."

"I guess I like to pity myself," whispered a woman on the other side of the room. "Yes. Okay. My name is June. I am a predator, and I consume a ton of self-pity every day." Looking sad, she checked the dreamers around her as if they might afford her a little of the pity she craved.

The woman standing beside her was engrossed in her own conversation, however. With her eyes closed and her hands folded in prayer, she looked as if she were confessing her sins to an invisible priest. "I am Monica," she said in reverent tones. "I am a predator, and on most days I'm hunting for approval. Today, most of all."

"My name is Tanya," said another. "I have lived for years on a diet of self-hatred." She hesitated, adding, "Forgive me; I didn't know."

Forgive me; I didn't know. Sarita sat in a straight-backed chair, viewing the exercise with fascination. Lala stood behind her, fingers tapping the back of the chair uneasily. They were in the twenty-first century now, and don Miguel was teaching again, walking around the room, listening to the confessions, and offering the presence of truth to those whose courage failed them.

"This is a misuse of ritual," huffed Lala. "Confession has always been a sacrament designed to purge the body of its sins."

"How well I know it," Sarita agreed. "It all seems logical and true, until we see that the body was never the sinner."

"How can you say—"

"Because the body is impeccable. The messenger is not. You, my dear, are not."

Lala clutched the chair until her hands turned white. She had no response, it seemed; but Sarita felt confident that she would think of something soon.

"Don't worry yourself," Sarita added. "You can be redeemed."
As she looked around the room, she had a realization. "Right
now, it seems you are being redeemed."

They both surveyed the room, listening to the sounds of
confession. Sarita marveled at the simplicity of a ritual whose
origins were as old as humanity. Confess. Admit that you've
sinned—that you have harmed your own human being in
some way. Repent—that is, make an agreement to stop. Do
penance, which means to offer yourself forgiveness. Modify your behavior from that moment forward. Sacraments do
not require a priest; justice is offered by the perpetrator of the
crime, whatever that crime may be. Confess, repent, forgive. It
is done.

Miguel had altered this dream noticeably. He was still the
teacher, the shaman, to most minds, but the message had
shifted, as had the discipline. He still took large groups to Teo.
He continued his Circle of Fire ceremonies, but now there was
this: a monthly workshop in San Diego, where students came
and worked for three days at a time. For one entire weekend,
they lived together, slept together, and learned together. They
listened, they shared, and they participated in exercises like
this. They laughed and they cried. They revealed their anger
and they howled with joy. As Sarita had done with her students,
they dreamed. Sitting in circles, one inside the other, and facing each other—hands on thighs, back straight—they closed
their eyes and allowed the mind to wander. This gave them the
chance to perceive themselves fully—as both the story and the
storyteller. They were the noise in their own head, and they
were the means by which the noise could be silenced.

Sarita smiled with recollection. He had taken the principles
of her teaching and expanded them. He was applying lessons
from the greatest stories ever told to challenge the minds of his
students. He was asking them to take responsibility for their

personal stories. Frightening or inspiring, all human storytelling came from the same author. The mind established the plot and the moral, and the human followed. These students were learning to modify their own tyranny over the body.

"This is stupidity," Lala grumbled, interrupting the old woman's thoughts. "What good can come of it?"

"For you, much good. For the human being that each of them sees reflected in the mirror, peace and salvation."

"Salvation? Salvation from what?"

"Once again, La Diosa, from you."

Astonishingly, tears welled in Lala's eyes. She turned away, but in every direction there were reflections of her own disquiet. Every face expressed a desire for atonement. Echoing off every wall were the sounds of admission and redemption. How could she turn away from this? *Forgive me; I didn't know.* The words moved her, as if she were speaking them to herself.

"My name is Amy," said another woman, standing close to her. "I am a predator. I eat self-rejection." She, too, was touched by emotion, but she continued to speak. "I want to love better, and I want this body to feel the blessings of my love."

As the two women watched, Miguel stepped toward her, closed her eyes with his hands, and blew hot breath into her mouth. *The kiss of the nagual.* Sarita smiled. The woman made a soft sound, a whimper, and let her body collapse gently onto the floor. There, mindless and surrendered to truth, she would start to comprehend.

Sarita looked at Miguel and saw how inspired he was. He had renewed his efforts. In his words, there was a message worth hearing. In his being, there was a necessary disturbance. She sighed, wondering how this adventure was to end. This scene, this memory, was one of the last. He would take one more power journey to Teotihuacan, and then his valiant heart would fail. What must she do now? What could she do to help him resist death? She had gathered so many memories, so many pieces of him, just as she had been instructed. What now?

She turned to Lala, who was facing the mirrors with an odd expression, as if seeing a reflection that was not there. The woman looked afraid, and there were tears shining on her beautiful face . . . a face that used to belong to Sarita. Yes, it was possible that she had looked like Lala once, when she was young and impassioned. She consoled herself with the thought that there was no resemblance now. She was not this woman; she was not the desperate voice within her own mind, forever clamoring to find lodging in the minds of others. She was not the tyrant or the harpy. She could not control her son as if he were an infant. He made his own decisions and accepted their consequences. It was possible that he remembered her as someone warm and welcoming, even as he slept, and that he honored her as his teacher; but even her most compelling words could not reach him now. She must lay down her illusions and be done with it.

"La Diosa," Sarita said respectfully, and the redhead turned to her. "This journey is reaching its end. Come with me, and we will give our final blessings to my son . . . where he lives now."

Lala hesitated, trying to think of something scornful to say in response. She'd said from the beginning that this was a stupid waste of time—and the old bird should be reminded of that—but this was not the moment. *Not now,* she warned herself. *Not here.* When she was feeling herself again, she would have a thing or two to say about all of this. Brushing away the damnable tears, she nodded and followed her companion out of the room.

I will linger here a little longer, breathing in the feel of the place and listening to forgotten conversations. I loved classrooms like this . . . and can still feel the heat generated from inspiration's fires. Toltec Dreaming, as I called my new curriculum, attracted new generations of students and encouraged them to begin a journey back to common sense. Looking into the eyes of my students and sensing their eagerness to learn, I'm reminded of the power and significance of those monthly workshops. They gave me happy opportunities to try different teaching methods, and they led to real transformation in the lives of my students. They wanted to know things, of course. The mind is built to process new knowledge, and eager to share it with other minds. The knowledge I gave them wasn't new, but it encouraged fresh perspectives. It inspired the kind of self-reflection that challenged them, and it pointed the way to unimaginable freedom. I wanted each of their minds to appreciate its intended power. I wanted every mind to see itself as the hero—on a mission to save the human.

The excitement that lived in this classroom heightened as realizations came, one after another. There were other emotional consequences, too, of course. Disturbances of the mind are opportunities for change, but the smallest disturbance is often unwelcome, even to those who insist that they're eager for it. Transformation is the result of big shifts in perceptions and responses. The process can be uncomfortable, since the mind is reluctant to notice itself, to take responsibility for its stories, and to change. When the mind is in acute resistance, the emotional body pays a price.

As I've mentioned, if certain students couldn't move beyond the fear and anger that came from making big changes, it was only sensible to let them find another way to awareness. It was unacceptable to let them use Miguel against themselves. I, too, felt the emotional consequences of letting go, but those students needed to develop more faith in themselves and less in me. I trusted life and encouraged all of them to do the same. It's because we respect others that we allow them to be the dreamers they want to be. It's because we love them that we let them make their own choices, whether we understand those choices or not.

Emma had left my dream before these new workshops began. She had decided to throw away the last crutch and continue her journey without me. She was a good warrior, facing her challenges alone and daring to venture into new dreams and unfamiliar situations. She may have felt she was on her own, but the bond between us was always strong. As I had done with Dhara, and would do over and over again with my sons, I let her go, and I waited.

Letting go of my outward connection to Emma may have been a painful choice, but it was made out of respect and the desire to be generous. There was a real possibility that we would never see each other again, an outcome that made me uneasy; but what gift could I offer that was more precious than her freedom? No one can become a master without practice, without waging a lonely war against the voice of knowledge and winning. If the war ever became too difficult for her, I counted on Emma to reach out to me. She would always be welcomed back into my dream, my arms. She had never left my heart.

Love would determine everything, as it always has. My heart is full, even now. My body, struggling as it is to stay in touch with the forces of life, is responding to every mental image and to all the small insinuations of excitement that rise from memory. Life is offering every resource for me to use in this struggle. Every person in my life is now an ally to my understanding.

Everything I've learned and experienced has become a servant to my survival. Everything that lives, that moves—everything that spins beyond knowing—is in play. Without thinking, I'm aware. I'm aware of the totality. I'm aware of what is and what once seemed to be. I'm aware of the messenger called Miguel, the message he delivered, and the love he shared.

With awareness, reality shifts. We're able to see that being alive is its own purpose, and that everything we do while we live is an expression of our art. How we think, converse, and behave determines the quality of this art. We are life, but we've been perceiving from the mind's point of view for so long that it's difficult to see beyond it. The mind denies its separateness when it says, "I feel good," or "I feel bad." Realistically, the mind doesn't feel; the body does. The mind tells a story about guilt, anger, or even embarrassment, and the body pays the consequences.

I asked my Toltec Dreamers to look into the mirror, into the eyes of the human being, and let the storyteller introduce itself. "My name is . . ." it begins.

If the mind can hear itself, it can make changes. Self-judgment may be among its favorite foods, but that appetite has to be recognized before it can be modified. The moment the mind identifies itself as a predator, it has committed to becoming an ally. The mind can create chronic problems for the human, but it can offer miraculous solutions as well. It has the ability to imagine. We routinely imagine conversations that haven't happened. We imagine alternate realities and phenomena yet to be discovered. We reimagine the past and fantasize about the future. We imagine gods and demons. We imagine horrors and wonders. Among the many things we imagine, demonic possession may be the most interesting and revealing. All cultures make reference to it. They may fear it or pass it off as a joke, but possession is real.

We know what it means to possess a thing. We buy something, we pay for it, we trade it for something else of value, and it is ours. Yes, we also imagine that we own people—the people we love, the children we raise, employees who receive a salary from us, and the land we occupy. In most cultures, possessing things is extremely important, and it's been an accepted practice in many cultures to possess other humans. What we may not see clearly is that *all* humans are possessed, whatever their beliefs and traditions. They are possessed by something intimate and eerily present. The human body is possessed by knowledge.

Years ago, I was speaking about this to a large audience. A woman in the crowd stood up to make a comment, and I sensed a deep and prevailing anger in her words. It seemed she was reacting to something she'd heard me say. I called her up onto the stage. I gave her a hug, which she stiffly accepted, and invited her to talk more about her life and her point of view, even encouraging her to release her anger—to get it all out. She began to talk, hesitantly at first, and then more easily. The more she talked, the more aggressive she became. I stood beside her, saying nothing, as she delivered a tirade against men, against God, against life. Her voice got louder, her body shook, and her face took on a demonic expression. Looking at the audience, I saw people's initial amusement turn to concern and then to shock. I saw fear in many eyes.

After a few minutes of this, the woman's anger began to dissipate and she started to cry. I gave her another hug, which this time she accepted gratefully and wholeheartedly. The audience applauded her, impressed by her show of raw emotion, but they remained uneasy. When she had returned to her seat and the crowd had calmed down, I asked them if they had any doubt that possession was real.

"Now can you understand what I'm saying?" I asked. Of

course they could. We can externalize all our biggest demons, we can scare ourselves with images of devils and ghouls—or we can look inward, listen, and recognize the voice of knowledge that lives and speaks to us on a constant basis. That voice can sound very scary. We are not accustomed to hearing ourselves, and so we rarely acknowledge the messages of selfishness and cruelty that we deliver on a daily basis. It's an act of self-love to pay attention to what we think, and to modify that internal conversation.

Everything in life is intended to serve life, and we can turn knowledge into the servant it was intended to be by listening and by changing. Having mastered the art of creating hell in our own minds, we can choose to become big angels, impeccable messengers. Every dream should be respected, but any dream can change when it's ready. A willing mind can offer the kind of freedom we all want and can hardly imagine.

If we recognize that knowledge, serving life, can also be the solution to human suffering, we can alter the way we interact with other humans. We can change the way we imagine ourselves. We can stop believing our own voice, *all* voices, but keep listening and keep learning. "I am a predator, and I have an appetite for fear . . . for guilt . . . for anger," is one way to start. For my students, this generally brought the problem home and within reach of a solution. Aware minds can become hunters who crave truth.

My best students are still hunters, but they feed on different foods now. They were one thing, and now they're something else. They are loyal to the human, creating realities based on gratitude and generosity. Where the mind used to control and to punish, now it serves. Where it may have behaved like a villain, now it's a hero. Where once it craved only poison, now it has a taste for nectar.

23

ARE YOU A BEE OR A FLY?"

"What?" the man asked, confused.

"Are you a bee . . . or are you a fly?"

"A bee, I guess. Who wants to be a fly?"

Don Miguel nodded and turned to another student. The class was large, with students almost filling the big studio space they occupied. Most were seated in a comfortable chair, and they crowded close to the teacher as he spoke to them from a small sofa in the front of the room.

"What are you," Miguel asked a woman taking notes, "a bee or a fly?"

"A bee!" the woman answered brightly, dropping her pen. She was suddenly attentive, eager to take part in the discussion.

"Are you sure?" Miguel asked her, leaning close and looking

directly into her eyes. "Are you sure you don't like to eat a little poo sometimes?"

"Poo?" The woman was shocked and apparently insulted. "Of course not!"

"You never gossip with your friends? You never judge other people?"

"Well . . ."

"You never agree with the judgments made against you? You never get offended? Your feelings don't get hurt?"

"I think I sometimes like to feel sorry for myself."

"Because . . . ?"

"I feel rejected, maybe . . . feeling people don't understand me."

"So you eat a little poo?"

"Maybe."

Miguel turned to an older woman. "Do you ever judge your own human?"

"I guess so, but I eat honey most of the time," she said.

"You tell your body how beautiful it is, right?"

"Sure, yes."

"You tell this human that she is the love of your life." Miguel waited, while the woman let the idea sink in.

"The love of . . . ?" She blushed, searching for something to say.

"Who deserves your love and loyalty more?" Miguel continued softly, as if he were in a private conversation with her. "This human was there when you were born, and she will be with you until the end." He gave the woman a compassionate look. "She submitted to you right away and has always obeyed you, no matter how many ways you've neglected and abused her. She is the one who listens to all your secret confessions and accepts you the way you are. She is your constant companion and friend. She is the love of your life."

"Yes, I see," responded the woman, blushing more.

"Who do you treat better, your cat or your human?" Miguel asked a man sitting in the back of the room.

"I have two dogs," he said, "and there's no contest. They get much better treatment." The class laughed. "Honestly, they sleep on goose-down comforters and eat organic dog food."

"You probably treat your dogs better," Miguel agreed, "but there will never be a pet more generous to you than your human, or more loyal. What about you?" he asked a small woman in the front row.

"I have a cat," she said.

"Are you a poo-eater?" The woman looked down at her notepad and shook her head. It seemed for a moment that she would cry.

Wanting to give her time, Miguel turned to the woman beside her. "Are you?" he asked.

"I'm afraid so," the second woman said right away. "I'm always afraid, it seems. I'm afraid now."

The small woman sitting beside her stifled a sob, and Miguel opened his arms to her. She put down her notepad and sat next to him, laying her head in his arms. He stroked her hair and looked around the room. There was a man sitting against the far wall, his arms folded tightly against his chest.

"What do you prefer," Miguel asked him. "Honey or poo?"

"I prefer honey," he said, "but I'm used to poo. I admit it."

"Are you a big judge?" Miguel smiled, seeing how the man was casting disapproving looks at the woman he was holding.

"No. It's just that everybody's an idiot but me." With that, the classroom rang with laughter again. The man relaxed a little and placed his hands in his lap.

"So . . . you're a fly?"

"A fly?" The man frowned a little, considering.

"You like eating poo, so you're a fly," Miguel explained. "If you were a bee, you'd be eating honey all the time. You'd be eating love, respect, joy . . . no? Am I wrong? You would say nice things to your human. You would watch the behavior of other people, notice how they hurt themselves, and you would feel compassion. You would show everyone respect, because you'd respect yourself so much." He waited, observing faces all around the room. "Am I right?"

There was a rolling wave of agreement around the class.

"Can't flies learn to eat honey?" the man asked.

"You want to be a fly who eats honey? Go ahead!" Miguel grinned at him. "Just remember that, sooner or later, you won't be able to resist poo. You'll want your normal food—normal for a fly, of course."

"But if you weren't born a bee?"

"We were all born bees. Well, naturally, we were born humans. You understand that these are metaphors," he said, smiling. "We were born humans; and humans are born without a mental program. That comes later, when the child masters a language and begins to think. At that moment, the human becomes submissive to the mind. Thoughts decide what is real."

"What may be real," someone corrected.

"The mind determines what is real," he asserted.

Miguel looked down at the woman now leaning against his knee. He touched her cheek with his bunched fingertips, kissing, kissing. She seemed unable to respond.

"Even with thoughts in our head," he said, "even after we mastered the symbols, it took us a long time to learn to eat poo. As you've certainly seen, little kids are happy. They are naturally kind, and they love to be loved. They feel every emotion, but when they're very young, emotions come and go quickly. If they feel hurt, they get over it. If they feel afraid, they grow out of it. They're curious: they want to know everything about everything. They learn the names of objects, learn a language, and . . . vroom! A million opinions go into their heads."

Miguel looked around at his students, noting their attentiveness. "All those fears and hatreds and criticisms have a big influence on children," he continued. "Their taste for honey turns into a taste for the kind of food everyone else is eating. It isn't natural, but it looks normal. Everyone is doing it. They eat poison, and the body reacts emotionally. They hear mean, horrible things, and they can feel their bodies produce anger. They can say, 'Well, other people are eating anger, and they really like it, so I guess I'll eat it, too.' So anger has to be produced over and over, to keep them fed. Some people like guilt . . . *mmm*. It tastes so good after years of eating nothing else. Some adults still like to eat fear. It looks yummy, too, sometimes. Maybe it looks like shyness, nothing bad, so they make an effort and get used to the taste. To appear strong and in control, they eat their own fear in secret, and get fat on other people's fear. To look smart, they develop a taste for judgment. To be like grown-ups, to be like other humans, kids learn to eat poison until they have a tolerance. Poo begins to taste okay."

"That sounds terrible," the older woman said.

"It sounds logical," he replied. "Would it be better to say that the devil makes them do it? See how fear of the devil, and of God, turns people into poo-eaters—and they want to share all this poo with everyone else. They want to pay ten percent of their income to keep the devil away. They eat fear, guilt, shame, everything, and can't get enough. Isn't it better to see this, and to change yourself?"

"Change our eating habits?" someone asked.

"How about you first change your species?" Miguel suggested.

There was silence in the room—a silence so complete that the woman he held lifted her head, gazing around in bewilderment.

"You're not real," Miguel continued to the class, "but your human is. The human doesn't need to change, but you can. You can evolve . . . from a fly to a bee."

"By being aware," someone offered.

"Okay, let's put it another way," Miguel went on. "Let's say you acted like poo. Nothing you said was pleasant, all your thoughts were angry. You were poo. Who would be attracted to you . . . bees or flies?"

Everyone agreed that they would attract flies.

"And if you acted like honey, so sweet and so kind, always happy and respectful of yourself?"

The unanimous consensus was that they'd attract bees.

"Not exactly," Miguel said. "Something so sweet would attract *everything*." The group laughed. "No kidding," he insisted. "You would attract bees, flies, whatever. You need to be aware, of course. Awareness is the whole thing. It helps you see what you are, what you eat, and who is using you for food. In the world of animals, everything is eating everything else. The virtual world is a reflection, a copy. You can choose how you act in this world, and how you eat."

"I'm imagining the Serengeti, in Africa," one man said with a laugh.

"You are virtual. You can change your landscape. Why let your human feel it might be attacked by you, or eaten by you? Why allow that kind of irrational fear?"

"Good grief!" someone exclaimed.

"Your nature is to love; remember your nature. If you are acting against your nature, you're pretending to be another species. You've pretended so long that you actually are another species. Awareness means seeing, and remembering, the truth—it never went away. It never was gone, but your attention has been on the lies."

"I'm a bee," a young woman said, "who's been on a strict

· 354 ·

diet of poo so she can fit into society. I'm sick, I'm kind of disgusting—but I'm cool."

"In whose eyes?" Miguel asked when the laughter had died down.

"In the eyes of the flies."

"In the point of view of your adopted species . . . which is predator."

There was a resounding clap of approval from the back of the room as don Leonardo put his elegant hands together in delight. "Look at that!" he marveled. "Magical!" Leonardo glanced at the woman beside him. She looked disgruntled, as if she were ready to eat some of her own poison and nothing anyone said would soothe her raging appetite. He sighed, wishing he could have shared this moment with Sarita; but he had lost her in the tide of memories. She would have enjoyed this. Perhaps she would have learned from it.

"How do we know if we're reacting in the right way?" one student asked.

"How?" Miguel smiled. "If you're confused, ask yourself: What am I, and what am I eating? More specifically, are you a predator in this moment, or an ally? Are you eating honey right now, or are you eating poo?"

The room was vibrating with energy and intent. Don Leonardo was delighted. This, perhaps, was what his grandson would miss most. This was the place, these were the people, that he would be coming back to . . . if he came back at all.

"Someone will pick up where I left off; have no fear," a voice announced quietly.

Leonardo turned to discover Miguel, sitting in a canvas chair next to his grandfather, looking small and frail. This was not the teacher, talking to the class. This Miguel was pale and thin and still wearing the stained hospital gown. This one seemed to be grasping life with a weak, indifferent hand.

"They understand, and then they forget," Miguel said as he observed the class. "The distortion happens so fast."

"You give the gift of wisdom," Leonardo said, his eyes filled with comprehension, "and then you put the gift aside when you see that it is being corrupted. Later, you bring it back with another name. That, my boy, is the work of an agile teacher." Corruption happened easily and swiftly. He remembered this from his own days of teaching. Almost as soon as he put truth into words, his apprentices would twist the words into glib, comfortable lies. Miguel saw this clearly. His enthusiasm for teaching, his love of tactical solutions, could very well revive him now.

"You are kind, *abuelito,* and wonderfully tricky," his grandson said, smiling. "This memory won't bring me back, though. I did my best, and there's nothing more to be done."

"Perhaps your mother was right. You *are* as stubborn as a child."

"I'm as happy as a child, and looking forward to going home."

"You are no longer a child, Miguel, but a wise man and a master," Leonardo said, "so I will speak to you as one." The old gentleman shifted in his chair, looking directly into Miguel's eyes. "We who can speak *must* speak while we have the gift. We must love while we still live, and we must act while we still have the strength."

"I have no more strength to give this body."

"You have *all* the strength. Had I been such a shaman, I could have moved mountains. I could have moved Guadalajara to the ocean, where it belongs."

"I'm sure." Miguel laughed feebly. "But what would be the point?"

"When did you care about points and purposes, don Miguel?" his grandfather quipped. "Let Guadalajara stay where it is. Do you imagine all minds are like deserts, hostile to every seed?"

"I can't change an unwilling mind."

"You can bring your body back to health," Leonardo retorted. "Decide later if you care about human minds."

"I don't."

"Truly? Not the minds of your sons, and their future sons?"

"With or without a father, they are content now, and at peace."

"They are at war, *m'ijo*. This is a war they cannot end by themselves."

Miguel was silent, imagining José as he was now, wrestling with his own doubts. The boy had suffered too many knockout punches already. But what could he do for José now, when his own body was too weak to fight?

"You can pour yourself into the boy," said Leonardo, sensing an opportunity to win this argument. "You can give him all the wisdom of the *nagual*—leaving nothing of yourself, if you deem it necessary. Give him whatever you have left to give. Even now, he sits at home, inventing his own teaching out of sheer intuition, shouting to the walls like the inspired madman he was destined to be!" The old man adjusted his cravat and sat a little taller.

"I could do that, yes," said Miguel with a weak smile. "I could be there for him, as he is for me. And I could give all that I have."

"And the others? They need a father and a guide, too, although they may not see it now. They need you."

"My children," he murmured, conjuring an image of them in his mind. Would he really leave them as they were now, if he could choose differently? Don Leonardo was right, of course—he was a man of reason and imagination, after all. He could see that the boys deserved what any other apprentice deserved. He understood that they could use knowledge to develop a more intimate relationship with life. As aware humans they could learn to see more, see farther. They could use that enhanced vision to peer beyond the parameters of knowledge—and to jump. And yet the prospect of teaching that lesson again made Miguel weary. "I have no desire left for this world, you know," he said. "No lust, no passion."

"No lust? Don Eziquio would be horrified!" They both laughed, feeling an inexplicable relief. Leonardo laid a hand on his grandson's shoulder. It was still solid. He had lost some flesh, some muscle, but there was a vitality in him, waiting to catch fire. "Being a lover is part of the warrior's life, *maestro*. You would do well to consider that. A lover must stay alert and prepared for action."

Don Leonardo glanced back at the woman next to him, lowering his voice. These were matters he was loath to discuss in front of ladies. "A lover must be ready," he whispered urgently. "He must open his eyes, his ears, and all his senses to the smallest signs of dissatisfaction in his mate. His eyes must stay bright and his hands busy. He may not grow lazy or tepid; his blood must be ready to boil. Never rest! Never fail to notice a sigh, a warning, a flirtation. Answer her sighs, her carefully placed clues. Meet craving with craving. A woman must never suspect that you have become less able or less eager."

"Too late, *abuelito*." Miguel shrugged, his eyes sparkling with humor.

The old man changed his tone. "You have been distracted by the needs of your own body; I understand. Having died on several occasions, you—"

"Three occasions, at least."

"Having died on multiple occasions, you might assume that any lethargy on your part would be excusable, but—"

"It is excusable."

"What of the women you love? Indeed, you may ask, 'Why

should they want me now, now that I am half the man?' The answer—"

"Less than half."

"The answer is that they wanted you then!" the old man cried, with an expression that implied all of this was obvious. "They remember! Their hunger is sharp; their passions have not cooled. These women still call out for you in their sleep!"

"I'm not so sure."

"Don't be dull-witted! They call for you, but you are inattentive. You have lost your warrior's edge."

"There will be other lovers, better warriors. There's nothing left to fight, nothing to conquer."

"And nothing to redeem?" Leonardo asked. "Nothing, Miguel? *No one?*"

Miguel looked at his grandfather, noting the power of a thousand suns behind his cloudy eyes. He considered the old man's words, remembering an afternoon they'd spent together so many years ago. "The only conflict is between truth and lies," don Leonardo had told him that day. There was only one truth . . . but the lies we tell ourselves are countless. So what about Lala, the mistress of lies? Would he really leave her to her many indiscretions while he could still breathe? Would he neglect to plant a few more seeds of truth . . . if there was a choice?

The choice had been made for him, of course. He would come back, not because his family wanted it, but because life still pushed through him, taunting, insisting. Life was measuring his desire to collaborate again. One small action, born of desire, could spark a dying dream. Was it time for such an action? Looking into don Leonardo's eyes, it seemed to Miguel that his grandfather was asking the same question.

Miguel looked down at his body, scantily dressed and bruised by needles. He was reluctant to return to this broken man . . . but he needn't return with the same kind of force. He could walk lightly, serenely, through the human dream. He could offer comfort to the ones who wanted him, submissive to life's intent. He could be attentive to Lala, without yielding to her—Lala, the goddess he himself had once worshipped. She was expecting him, he knew. She was listening, looking . . . and craving the next tidal wave of love.

"I'm thinking of her," he said to his grandfather. Not the one with red hair and fiery eyes, he reflected, but the one called humanity . . . the one he'd known well and might come to know better, if he stayed.

The old man said nothing, waiting.

"Don't worry yourself, elder," Miguel said at last, taking his grandfather's hand. "I will be a guide for my students. I will be a father to my sons and a son to my mother. For women, I will be a lover in all the ways that I can."

"And . . . the other?" his grandfather asked, careful not to mention Lala's name.

"There is no other. She is all of them," he stated simply. "Of course I will be there for her." He gave his grandfather a warm smile and saw relief wash over the old man's face. If he failed, Miguel thought with a sigh, nothing would be lost. Women would still dream. They would yearn, with or without him. Men would strive. His sons would enjoy their days and nights. Life would go on . . . and, occasionally, a seed of truth would take root.

The two men sat in silence again, their attention eventually drifting back to the class. Unmoved by life's approaching tempest, Miguel stayed put, idly observing the dance between an inspired teacher and his eager, elated students. The smell of honey filled the room, and the sound of happy bees buzzed in their ears.

I TASTE NECTAR ON MY LIPS. I FEEL THE SWEET PULL of existence, and I hear life calling me to play. We are the sweetest of nectars. We are love. When we stop pretending we are anything else, we have regained our authenticity. We have found the thing we feared we'd lost.

We were brought into this world as authentic beings, and then we began to practice "who we are" until it became a mastery. At any stage in our adult lives we can unpractice; we can let go of our attachments to an identity . . . and we can *be*. An authentic life is a different sort of life. Unrehearsed behaviors lead to spontaneous interactions. A day spent with no expectations leads to unexpected wonders. A simple moment becomes an infinite moment, carrying with it the implications of total power.

Authenticity looks and feels different. When we are authentic, our presence is more important than our performance. We don't feel compelled to bring *me* into the room. Actions are taken, but actions and reactions aren't bound to what we believe about ourselves. We perceive, and we allow the human to respond to pure perception. We imagine, without boundaries. We see through all the senses, not just our eyes.

When I really began to see, it seemed I had an amazing advantage, but it also felt like an injustice: I was tempted to feel separate and misunderstood. Of course, both perceptions were supported by knowledge; so by ignoring my own doubts and opinions, I could surrender to the moment—that timeless reservoir of potential. I relaxed and accepted the truth of myself, however the truth was expressed. Everything is an event, one that is neither good nor bad. People are events. Ideas are events. I am an unfolding event; but most significantly, I am.

For many years, I had a strong sense of my relationship with the unconscious—that is, with total power. For a long time, I understood that I was matter and that I was also the force that moved matter. After my experience in the Great Pyramid of Giza, I had a strong sense-memory of that interaction. It continued to dominate my existence after I returned home, and I had trouble focusing on things that occurred around me. In the weeks after that journey, I was physically unsteady. I was seeing everything differently and acclimating to that change. The world of people and objects seemed increasingly distant to me.

I was still living within the body of a human being, however, and that body had obvious needs. It needed to eat, to sleep, to rest, to move, and to make love. If I was to live as a man, to continue affecting the human dream as I had been, I would first have to put my attention on the needs of my body. I would have to choose my priorities. Like someone setting out on a new journey, I would have to take one step, then another, and walk steadily toward another way of dreaming.

It was the beginning of a new century, and some of my students had been pleading with me to teach again. I agreed, with the stipulation that they resist superstition and fanaticism. I would begin regular workshops and focus on a limited number of students, my son José being one of them. Toltec Dreaming, my new curriculum, was a complete success. After the first year, I brought in more students, and the second year promised to be even more exciting. Students were traveling to San Diego from all over the country, even from Mexico and Europe. Old apprentices and beginners came together, and everyone was thrilled to be part of a new dreaming adventure.

At the beginning of our second year of Toltec Dreaming, Emma called me from her remote reality, asking for help. She had gone as far as she could alone and had continued using her own knowledge against herself. She was believing her own stories of injustice, and judging herself badly. In other words, her mind was not giving up its tyranny. She had taken many different actions—traveling, changing old patterns, and seeking new friendships—inevitably returning to the same place of unhappiness and frustration. She needed a savior.

Relieved to hear from her, no matter the reason, I urged her to come to a Toltec Dreaming workshop. She did. She understood the language of common sense, she saw what she could do to save herself, and she never left. She was under my wing again. At first reluctant to speak in front of people, she soon began putting her experiences into words. If she could imagine something, she could say it, and her observations ignited the imagination of other students. I pushed her constantly to teach, and she responded willingly, challenging her own beliefs and disregarding the judgments of those around her. She was on her way to becoming a dream master.

José was still unwilling to speak publicly, but he was listening and learning also. He was a natural dreamer, willing to expand his awareness in new directions and most comfortable within the world of his imagination. He was still young, with many years ahead of him—years that would provide endless opportunities to learn, to express himself, and to grow wise.

In February of 2002, I took all my students to Teotihuacan, as an end-of-year power journey. It wasn't exactly a graduation ceremony, but everyone deserved to feel that they'd accomplished something. My students had completed another year of personal challenges, and they were in good spirits; they were open to all experiences and possibilities.

Beyond that, I had seen an opportunity to change the dream again. It was my desire to shift authority to José so that he could gradually take my place. The time would soon come for him to be his own teacher and savior. It was important for him to build a different kind of dream from the one I had created, and so before the week in Teo was over, I named him as my successor. Not only was everything shifting again, but I had seen the inevitable conclusion to my human existence.

In recent months, I'd started to feel that I was here, but not here. I was becoming more in tune with the unconscious, looking toward existence from a vantage point that seemed far away. I sensed that I would leave my body soon. A physician might have told me that the heart was in bad condition, something that could be resolved through surgery. I was a physician, and of course I understand the truth of that; but whether I died of heart failure or something else, I knew I would die soon. We are all able to see with more than our eyes, and as we begin to see differently, we focus less on the things that our thinking tells us. Without knowing why, I saw my death.

I saw how life, pushing on this physical form, was being met with less and less resistance. The body is matter, and within my own body things were shifting fast. Submitting to total power, matter was preparing to disperse. This is how I saw what was happening. The soul, as I've so often reminded my students, is the life force that holds the particles of a universe together. Matter seems solid because of this unifying force. The soul makes it possible for every component to recognize itself as belonging to a particular universe. When matter can no longer support a dynamic interaction with life, its particles break apart and disperse. It seemed to me that I was reaching that point, and that the force holding my body together was preparing to release its grasp.

Life's process of reclaiming matter is a slow one that begins with the breakdown of the nervous system. The process continues to the major organs—the heart, the brain, and so on—until the body begins to decay and life finally consumes form. Soon my body would begin that process. Soon life would submit to life. I just didn't know when. I wanted to be ready, though, and I wanted to prepare everyone. José would take responsibility for my message in his own way. Emma would continue teaching. They were with me now.

Whether the other apprentices understood me or not, I gave them warning of what might come. Whether or not they chose to hear my words, I said what was necessary, and I released my attachment to any outcome. I told my students to love each other and to build dreams based on respect. I reminded them that my presence would be everywhere, because I was life. They were life. There was only life, and life's countless reflections.

Emma and I spent the last night of our Teo trip together, and the intensity of our lovemaking triggered a chain of events within my body. Matter seemed to meet total power, and the shock of it sent me soaring. My physical body reacted, and I had the overwhelming sensation of rising in love, away from the physical world and past the realm of human dreaming. I had no doubt that everything would change very soon.

At four o'clock one morning, shortly after our return home, I woke up to excruciating pain. My heart attack had begun.

◉ ◉ ◉

Is it not true, my angel of love, that the sun, the moon, and the stars have blended and are breathing in love with us? Is it not true, my angel of life, that in the divine order, you were born for me . . . and I was born for you?

The rising moon seemed to linger on the horizon, taking a moment to peer at the ancient ruins of Teotihuacan and flood the ghostly silence with light. An enormous shadow loomed from the base of the Pyramid of the Sun. It darkened the open plaza and the broad steps beyond. A thousand other shadows slipped through the city and moved along the Avenue of the Dead, crooked black lines crossing wide meadows of moonlight like greedy fingers snatching pieces of gold. It was nighttime. The world was asleep. Where there had been consciousness, there was none. Existence was fully forgotten and not yet imagined. Life was everything, the only thing; and matter remained motionless, dreamless, waiting to be touched again.

Not far from the silent pyramids, in a room at the Villas Arqueológicas, Miguel Ruiz lay sleeping. He was safely contained within the world of matter, wrapped in the arms of his woman, but infinity was beckoning him.

The room was completely dark. Eziquio could hardly see the bed, much less the couple who lay tangled in its sheets. Gandara stood beside his friend, squinting into the shadows and sensing a shift in the balance of things. The son of the son of don Eziquio had tipped the world over. He had crossed the imperceptible line. This was not a game, like most of the naughty tricks Gandara and his friends had played in their lifetime.

This was serious business. As Gandara looked more closely, he could see bright particles drifting away from the sleeping man in the bed—as if a membrane had been punctured and sparkling stardust was being released into all universes . . . streaming, scattering, and returning home.

"This may merely be the enchantment of dreaming," Gandara suggested suddenly, rippling the surface of depthless silence.

"This is the final disturbance," Eziquio said pensively. "He will be among us soon."

"You told me yourself, he is lying in a hospital—breathing, but not yet awake."

"How can he come awake?" Eziquio exclaimed. "You see what I see. Even now, he knows what we know. Very soon, there will be no human body to return to."

"My friend . . ." Gandara started to challenge him, and then hesitated. This voyage back to existence had left him touched and saddened. "Being human is harder than I remembered," he finished simply.

"I suppose," Eziquio said, nodding. "It certainly presents some challenging moments."

"Too many," said his friend, "in a space of time that is too short."

"We always knew this."

"I had remembered it differently. It was glamorous, I thought."

"The world of humans seems glamorous to dead men, *compadre*." He poked Gandara in the ribs. "We gave it glamor, no? When we were alive and full of passion!"

"I remember only the passion."

"The rest is unimportant."

"I see these two now, exhausted from passion, and it still seems difficult."

Don Eziquio turned to face him, noting that his friend had lost his good cheer. "Are you troubled by something, my friend?"

"I suppose fat old Gandara is yearning for home."

"We were called here, Gandara. Alive or dead, it is an honor to be needed."

"This has been a useful reminder. I was missing the lure of existence in much the same way a donkey misses the abusive hand of his master . . . or the way a whale, cruising weight-lessly through the ocean depths, yearns for—"

"Gravity."

"Exactly. We have been educated now. Shall we go?"

"And the shaman?" asked Eziquio, looking down at his great-grandson.

"The shaman is no longer a shaman," Gandara said point-edly. "Sarita's son is no longer Sarita's son. The good doctor no longer practices medicine. The lover no longer pines; the ally no longer exists. In summary, the man is no longer what he was. What would you have us do? Stop him—just as he is about to leap to freedom?"

Eziquio hesitated, acknowledging that fat old Gandara had a point. They were meddling in a world of thought and matter, the world Miguel had already left behind. He imagined poor Pedrito, injudiciously brought back from the dead, and quickly shook the vision away. They were no longer the idle pranksters, the mischief-makers. They were messengers now. They had been commissioned by life, were immune to the temptations of knowledge, and therefore were incorruptible—in theory.

"This is as far as we can go, my friend," he said with a sigh. "We would be ill-advised to serve the wrong master."

"What master? I serve no master," grumbled Gandara.

Eziquio scratched his stubbled cheek and became pensive again. In life, they had served their parents, their wives, and their community. They had served in the military, as had their sons. They had served God, perhaps unwittingly. They had served life by serving their own pleasures. Were not all humans in service to something? It was obvious that he, don Eziquio, was no longer just a man. It was also obvious that he was a part of the enduring dream of humanity, or he could not have been summoned. He existed in human imagination and in the stories told by the living. And now, in death, he served those stories.

It was not possible to be both the character that his son Leonardo imagined him to be and what his granddaughter thought he was at the same instant; but Leonardo and Sarita were not the masters of this dream. Everything in his dream was a version of the dreamer. The old man considered that for a moment. Had their meddling really made a difference? He and Gandara had leaped at the chance to tinker with a dying man's imagination, but the dream itself was in command. Through memory and intellect, it would use life's abundant information to revive itself and to heal. Ghostly, but loyal, he and Gandara were here to serve Miguel, the dreamer—while Miguel, in whatever way he could, was serving the ones he loved.

"What is there to do but serve?" he said, turning away. In that split second, the two men had left the darkened room and were standing in a small garden within the hotel courtyard. The night was cold, and the shadows were stretching, grasping. "I find it comforting to know—and significant—that my great-grandson did one thing impeccably," Eziquio said.

"And that is . . . ?"

"He lived the life of a man," Eziquio stated. "He ate, drank, blundered, and bled, like everyone else! He made love like a hero, fought like a warrior, slew dragons—and remained faceless. He moved among the sleepwalkers, awakening the few, the thousands. He laughed!" Eziquio's legs did a spidery jig. "He danced!"

"He had fun, indeed!"

"He was an agile artist, commissioned by life," added Eziquio.

Gandara looked at his friend in the strange light and nodded. "You are wise, old thorn, and a champion among men," he said with renewed cheer. "I forgive you for everything that came before."

"For what? I am blameless!"

"You? Blameless?" Gandara hooted. "Can you truly say that, after all you—" Gandara's eyes flashed defiantly and then softened. "Ah, rogue! You almost pulled me back again," he chided, chuckling lightly.

They fell silent, reflecting. Words and their mischievous masters belonged to another dream, another age. It was time for these honored guests to leave the human carnival. They exchanged amicable looks and strolled out of the garden, keeping in step with each other as hungry shadows slipped back into darkness.

"Wait! What of the mother?" Gandara asked, remembering Sarita. His friend stopped short, and together they listened to the sounds of the night.

"I no longer feel my granddaughter calling me," Eziquio said, trying to sense what would not make itself seen. He reached out to touch a red hibiscus blossom, folded neatly in its sleep. It was as smooth as velvet, pulsing under his fingertips and ready to burst open at the first flash of sunlight. Life was taking a breath, it seemed. Life was lying in wait.

"Perhaps Sarita has come to the same conclusion," Gandara surmised, "and taken her leave of this dream."

"Could it be possible?" Eziquio questioned.

"If we can do it, she can do it, my good friend . . . my true friend." Gandara smiled broadly, feeling a weight lift from his beefy shoulders. He picked one of the blossoms.

"True, you say?" Eziquio teased. "You and I were pure fiction—the figment of an ancient dream, the stuff of lies."

"But what amicable lies we were!"

"Eziquio, and his wondrous ways!"

"Gandara, and his beguiling . . . belly!"

"And let us not forget our friend Nacho—"

"—*and his daring band of tricksters!*"

The two men embraced, punching each other affectionately, as a full moon rose above the villas, the stone ruins, and the sleeping world. They laughed with unreserved pleasure. They laughed all the way back to the bright little spaces of memory they had come from—far from the jealous whimsies of the human dream—their footsteps falling in unison until they could be heard no more.

Sarita and Lala watched the two specters evaporate into the night. The women stood under a tree, just outside the room where Miguel slept.

"Let them go," Sarita said, sensing that Lala was about to call out.

"Oh, I will, and gladly," the redhead responded.

"It seems we have been here before," Sarita whispered.

"Hmm. Observing two lovers in bed." Lala snorted, looking up at the bedroom window. "The more things change . . ."

"He is not here," said the old woman, tugging at her shawl.

"Certainly he is—just as certainly as we are here."

"No. Where he should be, there is no warmth. He slips farther from the world of living things . . . and from me."

"Then call him back, old mother," Lala said, tossing her hair in a gesture of impatience. "He dares not run from you."

He need not run, Sarita thought silently. She had come to bid him farewell, not to be the pest in the garden, spitting venom and whispering lies. She wanted to leave this place of illusions and go to his bedside, where she belonged. She wanted to touch his very real hands, to kiss his cheek—and to wish him goodnight. She wanted to love and comfort him, as mothers do, one last time.

"Why have I insisted?" she asked aloud. "Why have I challenged him, quarreled with him, and clung so tightly to the past?"

Lala stared at her, open-mouthed. The answer seemed self-explanatory.

"I've done all that I know how to do," Sarita continued, "and all that I can. I've tried everything, and I have to stop . . . trying." Remembering words from another dream, she said, "Warriors, after all, die—"

"Warriors will die trying!" the other snapped. "They must!"

The old *curandera* looked at the vision of a woman beside her. They were more alike than she'd wanted to admit, but this is where the resemblance must end.

"I have done a mother's duty," Sarita said, "and more than most mothers could hope to do. I have recited the prayers, performed the rituals. I have summoned all my faith and the faith of my family. I have called on the saints and the ancestors, spoken to God and argued with Jesus. I have said the right words, believing in their power." She hesitated, looking at Lala. "I have played your games, as I always have."

"And always will."

"No," she responded swiftly, her eyes looking up at the window. "Miguel has helped me see things as they are."

Lala wanted to laugh, but the old woman's expression prevented her. It seemed as if Sarita had just grasped a divine secret. Her smile glowed softly in the moonlight. Her attention had wandered, and she was peering through the tree branches that lent shadows to the nighttime garden. Sarita was gazing past the stuccoed walls of the hotel, as if she could actually see the ruins beyond. She might have been walking by the pyramids, letting her imagination take her where inspiration had given birth to breathtaking form and where darkness played forever among flecks of light.

"I can let go," Sarita said, as if in a reverie. "I can relinquish . . . everything."

"You most certainly cannot," Lala stated flatly, sensing mutiny.

"I have seen so much since my search began," Sarita said simply. "I have seen the things I fully expected to see—familiar things, familiar people—but I have witnessed them all with different eyes."

"Of course," Lala said smugly.

"I have seen myself differently as well," Sarita said. "I have seen myself in you."

"It is gratifying to hear you say that," Lala answered cautiously. "And . . . so? How do you look . . . *in me*?"

"I look afraid and selfish," Mother Sarita said, softening her words with a smile. "I seem arrogant—a woman who sees only what she wants to see."

"My dear—"

"I am a woman who listens but will not hear," Sarita continued. "In you, I am truth's adversary and love's denial."

"You talk as if—"

"In you, I am a poor reflection of myself, *señora*," she concluded. "It is time for us to acknowledge our differences."

"We acknowledge nothing without my consent."

"Then I will say goodbye."

"Goodbye?" scoffed Lala, although she felt a jolt of apprehension. "You cannot end something where it did not begin!"

Surveying the little garden with theatrical disdain, she said, "This was not where we began the journey, sweet sister."

"We began in a darkened room, with two lovers." Sarita indicated the hotel room window above. Her eyes lingered there, as she wondered whether she would ever see her beloved son again.

"We began where I live, woman, and that is where we will return!" cried Lala, chin held high in indignation. "I will not lose you both!"

Even as she spoke, Lala wondered at her words. What was she saying? What did she hope to prove? There were other minds to play with. There would always be humans to taunt, to tempt, and to command. She was the thing they craved. If she offered them poison, they drank it. If she brought them the antidote, they took it. She defined good and bad, right and wrong. She delivered heaven and hell. She was every word, every secret thought. She alone decided what was real, for she was the empress of human dreaming. She was La Diosa: *la última diosa!*

As quickly as the words resonated with her, her anger dissolved into emptiness. The moon went dark and the garden vanished. The hotel evaporated, along with its hundred occupants and all their captivating stories. The ancient ruins, whispering wisdom's secrets among the crooked shadows, melted into obscurity.

Lala peered into vaporous space and saw nothing. She tried to call out, but her voice failed her. The old woman was gone, Lala spoke no more, and the visions were over.

25

S ARITA! WAKE UP!"

Jaime was watching his mother's face with concern, looking for signs of consciousness. On the doctors' counsel, Miguel's family had decided to take him off life support the day before. Today, Miguel was breathing on his own, but no one knew if he'd regain consciousness. In any case, nine weeks was enough. He would come back to them or not, as life decided. Sarita hadn't approved, of course. When Jaime had left her the day before, she'd been fretful. What he'd found this morning was an old woman curled on the floor of her bedroom without a blanket, unconscious and almost lifeless. Her breathing was slight and shallow, and her body was cold. He touched her cheek and drew his hand back in horror.

"Sarita!" he called to her, turning her onto her back. Her skin was ashen. Her eyelids were shut, and yet . . . there was the faintest hint of a smile on her lips. "Sarita!" he insisted, shaking her shoulders gently with both hands. She stirred.

"Stop . . . trying," she muttered softly.

"*Mamá?*" Jaime spoke the word tenderly and with wonder. His grip tightened on her shoulders. "Sarita?" She must have heard, for she said something else, something he couldn't comprehend. Little by little, he assured himself, she was returning to her senses. He whispered a prayer of thanks, spoke her name again.

"Sarita," he said, grabbing her icy hands. "*Mamá,* it's Jaime. It's morning now. You've done enough. You've tried everything—"

"Yes," she answered him weakly. She felt his hands rubbing life back into her fingers. His love spread warmth slowly throughout her body. As she took a ragged breath, her mind balanced itself precariously between worlds. "My son," she murmured, and her mind found its place in human reasoning. Her eyes came open. She looked into Jaime's worried face, recognized him, and smiled weakly.

"Take me to Miguel," she said quite clearly.

Miguel Ruiz was in a hospital room, miles away from the house where his mother had been dreaming. His first flash of consciousness came and then left again abruptly, as his body struggled to come awake. Light seeped into his brain, crying for attention. Sound roared in his ears. His throat was dry and raw. There were no tubes in him, nothing to help him breathe, but pain was moving through his body like wildfire. A cartoon show was playing on the television set. At the sound of angry voices and frantic music, his first impulse was to be afraid. Before his senses could accept the world he had once occupied, he was reacting to the noise of it, the startling violence of it. He'd expected this but could not have predicted its intensity. Did he have no defenses? Was he an infant again? Had the whole dream started over?

Someone was in the room with him—one of his brothers, perhaps. He remembered his brothers, the tyrants of his youth and the protectors. He remembered his parents, wise and always comforting. He had been blessed with so much love, even within the cold fury of the human dream. The noise booming from the television made him uneasy, as the force of the cyclone again roared, deafening him to truth. This is what humans lived with every instant—the mind's thunder and the primal currents of fear. He had played in the eye of that storm once; but now, coming back to this weakened body, he was frightened.

"Lala," he mumbled, his lips cracked and his voice frail. Had she returned to command the cyclone? He shifted his attention, chasing a dream that had just recently faded. With cartoon characters screaming in the distance, he reached back toward sleep and called forth images that had been slipping from memory: There was a wise man in one dream, a hierophant speaking to him from the tombs of human history. Miguel felt as if he'd died in such a tomb, and come awake again with a new awareness. There was a council of elders, all of them competing for titles, all of them fighting for attention. There were men who wanted to change him, and women eager to own him. There were other creatures—eagles, dogs, and demons. There were the great warriors of another era, watching from the shadows. There were tall and silent angels in his dream, and a young girl wanting to take him back to innocence.

His mother was always present, sometimes here and sometimes there, appearing in various shapes. She advised, she implored, but he couldn't recall her words. He couldn't see her face or sense her motives. He had visited the streets of

Teotihuacan where his ancestors had walked. He had heard them speaking in hushed voices, reciting parables from a forgotten time. He could sense their intent as he walked dreamlike up the Avenue of the Dead and climbed the Pyramid of the Sun. He watched himself as a younger man—teaching, creating, and honoring the masters. He had been a warrior, but it seemed the war was long over . . . and already won.

It hardly mattered who he was, then or now, or where he was being taken. He was in love, forever in love, and there could be no doubt. He blinked into the sunlight that shone onto him from a window, and dreams fled from his imagination like butterflies in a spring gale. Suddenly, he was back in bed, back in his hospital room. Miguel opened his eyes. On the television set, a cartoon mouse was beating a cat nearly to death. He winced and looked away.

"My angel," a woman said.

The voice, low and comforting, reminded him of someone who had been close to him once. He turned toward her, intrigued, and saw a face that was both kind and comfortably familiar. Memories showered over him like a million shreds of confetti. He was vaguely aware of events, as if he could hear them being discussed in the adjoining room, over the roar of the television. Apparent strangers were reciting testimonials about him. They were exchanging opinions, theories. He wanted to stop the conversation. *I know this man better than you!* he wanted to shout, but he wasn't sure it was true. It was hard to tell whose life was being represented in all the bits and random pieces of sound and light. Miguel strained to make sense of the pictures, the small traces of stories that sketched a human life, but it seemed of little use. He gave up. What he could not remember, others would. Everyone was sure to provide a commentary. Everyone would be eager to talk, and no one would remember to listen. He knew where he was now.

Miguel took his first deep breath and felt his body relax. What other signs did he need? The game was on. This was life, pushing him and moving him. He felt love revive him and wash away the fear. He blinked, and blinked again. The woman beside him squeezed his hand, her eyes flooding with happy tears. She was old, but life played a melody through her that still rang clear and true. Somewhere in her smile he recognized a friend, an ally. He imagined smiling back at her, and in seconds a grin spread across his features. The woman responded.

"I am with you, my son," she said softly, comprehension sparkling behind her weary eyes. Sarita reached out to touch his face with cool, smooth hands, infusing her words with significance. "I'm with you now."

⊛ ⊛ ⊛

Is it not true, my angel of death, that in this corner of the world everything is more beautiful and love is magnificent?

Lala was standing in the shadow of the Tree of Knowledge, where Mother Sarita had first discovered her. Everything was very much as she had left it, and she was alone, just as before. Bleak storm clouds threatened the sky, and bursts of lightning illuminated the dreamscape. Earth floated belly-up in the vast sea of space, and the majestic Tree of Life towered in the distance. She looked toward the other tree, but could see nothing there. Oh, it blossomed. It held every sweet fruit imaginable, but no man sat upon its undulating branches—not even a small one, oddly dressed.

What had gone wrong? Had she pushed too hard, or not hard enough? She had followed the mother, indulged countless whims, and asserted her own will when she could. She had done all she knew how to do, but he was gone. He had made a choice: he had returned to the world of consequence. This time, he had no doubt. He had visited her world briefly, teasing her with invitations and plaguing her with hope. Admittedly, she had been hopeful at first, smelling Sarita's fear. She had hoped, and she had indulged that hope with vigorous certainty. It was hard to explain what had happened then. One memory led to another, and now mother and son were together. He was no longer curious. He was no longer hers.

Well, what did it matter? There were humans being born every day. Not all of them would be so curious, of course—looking past the reflection to seek the truth beyond; but wherever there was a question, she could provide an answer. *Why?* people loved to ask. *Who am I? How am I? What is my purpose, my future, my fate? Where is right? When is wrong?* Any kind of art was easy when the canvas cried out so desperately to be painted.

A leaf fell from her own tree. It floated among flecks of shade, danced in downward spirals, and lightly tapped the ground. She touched it with her foot and it crackled. Another one drifted past her face, and another, and soon leaves were gathering around her like dying soldiers. Dry, drifting leaves. Much had changed in her absence. What had happened to this landscape of perfect symbols? She pulled an apple from a nearby branch and bit into it with fierce anticipation. It tasted bitter now. She threw it to the ground and contemplated the brilliant, alluring tree in the distance.

The Tree of Life returned her gaze as it rose resplendently toward the sun. With a thousand branches stretched wide, it seemed to be welcoming her into its arms. Touched by the sight of it, she wondered how she must look from its serene perspective. She experienced a vision of herself in that moment as something indistinct—as a murky cloud waiting to transmute into breathable air and become attainable to light. She was a mirage, peering through her own haze to catch a glimpse of something real. She was a fanciful thing, a slight and slender imprint upon an infinite backdrop.

This was the revelation that she most resisted—a message often delivered, but never heeded. The shaman was not the only one who dared to educate her, nor would he be the last. He was not the first to subject her to love's ruthless scourge, and he was not the only visionary to predict her redemption. She could imagine him now, playing his silly games until they no longer amused, and then gleefully upending agreements. With every change, he created a new dream, inviting the world to come in and play. Was she not a game-master as well, or had her rules become too rigid, the game too perplexing?

Lala took a breath of heavy, stultifying air. The tree beside her seemed to be receding into its own massive shadows as her attention remained on life, a thing she had never trusted and hardly considered. How would it feel, she wondered, to surrender to something that could not be measured or understood? What might a world look like where truth had no adversary and its mystery was met with no resistance?

What was truth, she mused, but the lingering silence at the end of a valiantly posed question? Sitting uneasily within that silence, she felt a wave of emptiness swallow her words and leave her reeling with sensation. All was sensation. What seemed to have existed once, existed no more. Gazing around her, she saw that both trees had vanished from the timeless landscape. The planet was also gone from view, taking with it the flickering lights of human dreams. There were no symbols left—none but the stunning likeness of a woman, a priestess without an altar, standing windswept and alone in a make-believe world.

In that instant, the fantastical sky heaved and wept. Rain showered upon her and upon the dry and dusty terrain beneath her feet. Wind rushed toward the piles of dead, cracked leaves and sent them scattering. As one leaf scuttled by, Lala bent down to retrieve it. It was now only a showpiece of curling points and withered veins. She held it in her hand, watching raindrops tap its brittle surface. At their touch, the leaf seemed to take a quick breath and come to life. Lala recalled the infant boy set abruptly on a metal table, left to take life's first gasp on his own . . . and a flood of feeling washed through her. As the little leaf devoured each drop, its flesh began to soften and to green. Its stem strengthened. Its pointed tips trembled then, reaching hungrily for the unseen sun. The sun then appeared, shooting through clouds and striking them both with flawless light.

Lala uttered a cry of surprise as life surged through her to reclaim its finest reflection. The exquisite and enduring illusionist was made new in that moment. Her mood turned calm. Her eyes turned clear. Moving her senses outward, beyond the limits of knowing, she was stunned to feel the unfaltering pulse of life. Could life sense her as well? Could life see her, hear her? Did language matter here? She paused, searching for words immaculate with intent, and found her voice in the midst of mystery.

"I am with you now," she said.

Reader's Guide

Preface and Prologue

Q1: In the beginning of the book, Sarita enters the dream world on a quest to restore her son Miguel to life. Do you often interpret your sleeping dreams? How is that different from commenting on your waking life?

Q2: Miguel welcomes death with the gratitude of a warrior who fought well and wishes for a safe homecoming. What are your feelings about death?

Chapters 1–5

Q3: When Sarita tries to persuade Miguel to return to his body, we learn that he was once a shaman. Are you familiar with shamanism? What do you think is a shaman's unique skill?

Q4: In this book's narrative, why does it seem that Lala sometimes resembles the one looking at her? Why does she appear to dislike the smells and the chaos of life?

Q5: In the book, Miguel remembers his childhood interactions with girls, and ponders how they taught him that seduction is vital to life, that suggestion provokes imagination, and that imagination builds reality. How do you think your childhood encounters with the opposite sex still affect your relationships today?

Q6: An angel is a messenger. Miguel tells us that "it is an unusual messenger who uses seductions of the mind to benefit another human being. It is an iconic messenger who applies this skill to benefit humanity as a whole." What kind of messenger do you imagine yourself to be?

Q7: Miguel's grandfather tells him: "All the things you've learned in school, and everything you think you understand about life, comes from knowledge. It isn't *truth*." Can you see how knowledge can be perceived as a reflection of truth, and how that reflection (the sum of your opinions and beliefs) is a distorted version of truth?

Q8: Sarita insists on having her son returned to her, but Miguel tells us, "For her troubles, she [Sarita] will bring home a pretender—the flesh-and-blood likeness of her youngest son, who has already found the truth, and has gleefully dissolved into its wonders." In this story, how will Miguel be a pretender if he returns to his life? Have you ever felt like an actor in someone else's play? Has a life-changing event ever made it difficult to return to your normal routines?

Q9: Sarita is helped in her quest in the dream world by her father and grandfather, both long dead. Do you ever converse with loved ones who have died? What is your relationship to the people you have lost?

Chapters 6–10

Q10: Ancient Toltec spiritual warriors allowed themselves to be consumed by a metaphorical snake, in order to emerge reborn as aware beings and to master death. What do the words "mastering death" mean to you?

Q11: When Miguel meets Dhara, he senses that they will transform each other's lives. Have you ever met individuals you sensed would change your life? Did they? What part did you play in any changes and transformations?

Q12: Can you feel the activities of your mind as separate from your body? How do your thoughts affect the body, emotionally and physically? When you alter your thinking, do you then feel a different emotional result?

Q13: Miguel recalls a car accident as the beginning of a life-change for him. What, if any, traumatic experiences in your life have offered you an opportunity to reevaluate things? Did you change any part of your reality as a result? Did your personality change? Was there wisdom gained, and how did that wisdom become manifest in your actions?

Q14: In this story, hell is described as a marketplace, the *mitote* of ideas in our heads. Do you sometimes feel the kind of confusion that comes from too much thinking? If you want relief from the noise, how do you usually accomplish this?

Q15: Do you think it's true that humans are addicted to suffering? To what degree do you make yourself suffer over ideas, people, or opinions about your life?

Chapters 11–15

Q16: Try writing the story of your life, and see which memories tempt you to feel emotional pain. How many times do you need to rewrite your story before those memories no longer distress you?

Q17: *Nagual* and *tonal* are words that describe infinite life and all its very finite manifestations. Allow yourself to experience yourself as the *nagual,* then the *tonal,* and then the bridge between the two.

Q18: In Miguel's view, *love* is synonymous with *truth.* Have you ever used love as an obstacle to truth in your life? Have you used it as an excuse to suffer?

Q19: Have you ever been able to love without conditions? How has being loved unconditionally at any point in your life helped you to be more authentic and confident?

Q20: Black magic is the art of self-defeat. When, in the course of your life, do you remember using black magic on yourself? Do you still?

Q21: In your experience, have changing perceptions led to personal transformations? Have you ever deliberately changed a belief or quit a habit? Did the change lead to other changes?

Chapters 16–20

Q22: Can you see how everyone in the human dream is competing for others' attention? Can you see how they may be unaware of the amazing power of their own attention?

Personal beliefs typically control our attention. How would it help your life to take charge of your own attention?

Q23: Awareness means seeing what is, without judgment. Can you use the memories of your life to create a clear awareness of yourself in this moment?

Q24: A *toltec* is an artist; the ancient Toltec masters were artists of life. How is your own life a work of art?

Chapters 21–25

Q25: Have you noticed how fanaticism can alter perception and corrupt behavior? When, in your experience, have you been driven to a kind of fanaticism? How has obsession hurt you, do you think?

Q26: Near the end of the book, the reality of death is explained in detail. How have your views on the subject changed as you've journeyed through these chapters?

Q27: In what ways are you practicing authenticity? In what ways do you still practice being what you're not?

Q28: In many ways, knowledge is made aware of itself in this story. Can you think of ways that you've become aware of yourself as the voice of knowledge—as both the tyrant and the savior in your own wonderful story?

Acknowledgments

I would like to express the most profound gratitude to my parents, Sarita and José Luis, who made it possible for me to exist in this body and to know myself as the eternal force of life. Their generosity and impeccable guidance gave me the confidence to love, to receive love, and to share my presence with the world.

I give thanks to my grandfather, don Leonardo, for his remarkable wisdom and for the lasting impression he made on my heart and my imagination.

I am continuously grateful to Barbara Emrys, my coauthor on this beautiful book, for her many contributions to my life. Over the past two decades, as her creative instincts have developed into genius, it has been my pleasure to witness the evolution of a gifted messenger and to receive the benefits of her enduring loyalty.

This book would not have been possible without the enthusiasm and support I received from HarperCollins Publishing and the entire team at HarperOne and HarperElixir—especially Michael Maudlin, Claudia Boutote, Mark Tauber, Melinda Mullins, Kim Dayman, Terri Leonard, Adrian Morgan, Natalie

Blachere, Libby Edelson, and Josey Gist. I offer my full appreciation for the respect they have shown me, my family, and my staff, and I look forward to our continued collaboration.

Lastly, I give heartfelt thanks to my readers, whose desire to change their world—and to kindle a new relationship with truth—will always be my greatest reward.

About the Authors

Don Miguel Ruiz is the international bestselling author of *The Four Agreements* (a *New York Times* bestseller for over seven years), *The Mastery of Love,* and *The Voice of Knowledge,* and co-author of *The Fifth Agreement.* His books have sold over seven million copies in the United States and have been translated into dozens of languages worldwide. He has dedicated his life to sharing the wisdom of the ancient Toltec culture through his books, lectures, and journeys to sacred sites around the world.

For information about current programs offered by don Miguel Ruiz and his apprentices, please visit www.miguelruiz .com.

Barbara Emrys is an inspirational teacher and the author of *The Red Clay of Burundi: Finding God, the Music, and Me.*

WITHDRAWN